T0215199

Advances in Intelligent Systems and Computing

Volume 395

Series editor

Janusz Kacprzyk, Polish Academy of Sciences, Warsaw, Poland
e-mail: kacprzyk@ibspan.waw.pl

About this Series

The series "Advances in Intelligent Systems and Computing" contains publications on theory, applications, and design methods of Intelligent Systems and Intelligent Computing. Virtually all disciplines such as engineering, natural sciences, computer and information science, ICT, economics, business, e-commerce, environment, healthcare, life science are covered. The list of topics spans all the areas of modern intelligent systems and computing.

The publications within "Advances in Intelligent Systems and Computing" are primarily textbooks and proceedings of important conferences, symposia and congresses. They cover significant recent developments in the field, both of a foundational and applicable character. An important characteristic feature of the series is the short publication time and world-wide distribution. This permits a rapid and broad dissemination of research results.

More information about this series at http://www.springer.com/series/11156

Rituparna Chaki · Agostino Cortesi
Khalid Saeed · Nabendu Chaki
Editors

Advanced Computing and Systems for Security

Volume 1

 Springer

Editors
Rituparna Chaki
University of Calcutta
Kolkata, West Bengal
India

Agostino Cortesi
Università Ca' Foscari
Venice
Italy

Khalid Saeed
Faculty of Computer Science
Bialystok University of Technology
Bialystok
Poland

Nabendu Chaki
University of Calcutta
Kolkata, West Bengal
India

ISSN 2194-5357   ISSN 2194-5365   (electronic)
Advances in Intelligent Systems and Computing
ISBN 978-81-322-2648-2   ISBN 978-81-322-2650-5   (eBook)
DOI 10.1007/978-81-322-2650-5

Library of Congress Control Number: 2015951344

Springer New Delhi Heidelberg New York Dordrecht London

Printed on acid-free paper

Springer (India) Pvt. Ltd. is part of Springer Science+Business Media (www.springer.com)

Preface

The Second International Doctoral Symposium on Applied Computation and Security Systems (ACSS 2015) took place during May 23–25, 2015 in Kolkata, India. The University of Calcutta collaborated with Ca' Foscari University of Venice, Bialystok University of Technology, and AGH University of Science and Technology, Poland, to make ACSS 2015 a grand success.

The symposium aimed to motivate Ph.D. students to present and discuss their research works to produce innovative outcomes. ACSS 2015 invited researchers working in the domains of Computer Vision & Signal Processing, Biometrics-based Authentication, Machine Intelligence, Algorithms, Natural Language Processing, Security, Remote Healthcare, Distributed Systems, Embedded Systems, Software Engineering, Cloud Computing & Service Science, Big Data, and Data Mining to interact.

By this year, the post-conference book series are indexed by ISI Compendex. The sincere effort of the program committee members coupled with ISI indexing has drawn a large number of high-quality submissions from scholars all over India and abroad. A thorough double-blind review process was carried out by the PC members and by external reviewers. While reviewing the papers, reviewers mainly looked at the novelty of the contributions, at the technical content, at the organization, and at the clarity of presentation. The entire process of paper submission, review, and acceptance process was done electronically. Due to the sincere efforts of the Technical Program Committee and the Organizing Committee members, the symposium resulted in a suite of strong technical paper presentations followed by effective discussions and suggestions for improvement of each researcher.

The Technical Program Committee for the symposium selected only 37 papers for publication out of 92 submissions. During each session, the authors of each presented paper were given a list of constructive suggestions in a bid to improve upon their work. Each author had to incorporate the changes in the final version of the paper as suggested by the reviewers and the respective session chairs. The symposium Proceedings are organized as a collection of papers, on a session-wise basis.

We take this opportunity to thank all the members of the Technical Program Committee and the external reviewers for their excellent and time-bound review works. We thank all the sponsors who have come forward toward the organization of this symposium. These include Tata Consultancy Services (TCS), Springer India, ACM India, M/s Business Brio, and M/s Enixs. We appreciate the initiative and support from Mr. Aninda Bose and his colleagues in Springer for their strong support toward publishing this post-symposium book in the series "Advances in Intelligent Systems and Computing." Last but not least, we thank all the authors without whom the symposium would not have reached this standard.

On behalf of the editorial team of ACSS 2015, we sincerely hope that this book will be beneficial to all its readers and motivate them toward further research.

Rituparna Chaki
Agostino Cortesi
Khalid Saeed
Nabendu Chaki

Contents

About the Editors

Rituparna Chaki has been an Associate Professor in the A.K. Choudhury School of Information Technology, University of Calcutta, India, since June 2013. She joined academia as faculty member in the West Bengal University of Technology in 2005. Before that she served the Government of India in maintaining industrial production databases. Rituparna received her Ph.D. from Jadavpur University in 2002. She has been associated with organizing many conferences in India and abroad by serving as Program Chair, OC Chair, or as member of Technical Program Committee. She has published more than 60 research papers in reputed journals and peer-reviewed conference proceedings. Her research interest is primarily in ad hoc networking and its security. She is a professional member of IEEE and ACM.

Agostino Cortesi received his Ph.D. degree in Applied Mathematics and Informatics from University of Padova, Italy, in 1992. After receiving a postdoc at Brown University, in the US, he joined the Ca' Foscari University of Venice. In 2002, he was promoted to full Professor of Computer Science. In the recent past, he served as Dean of the Computer Science program, as Department Chair, and as Vice-Rector of Ca' Foscari University for quality assessment and institutional affairs. His main research interests concern programming languages theory, software engineering, and static analysis techniques, with particular emphasis on security applications. He has published over 100 papers in high-level international journals and proceedings of international conferences. His h-index is 15 according to Scopus, and 23 according to Google Scholar. Agostino served several times as a member (or chair) of program committees of international conferences (e.g., SAS, VMCAI, CSF, CISIM, ACM SAC) and he is on the editorial boards of journals such as "Computer Languages, Systems and Structures," and "Journal of Universal Computer Science."

Khalid Saeed received the B.Sc. degree in Electrical and Electronics Engineering from Baghdad University in 1976, and the M.Sc. and Ph.D. degrees from Wroclaw University of Technology, in Poland in 1978 and 1981, respectively. He received his D.Sc. degree (Habilitation) in Computer Science from Polish Academy of Sciences in Warsaw in 2007. He is a Professor of Computer Science at AGH

University of Science and Technology in Poland. He has published more than 200 publications—edited 23 books, journals and conference proceedings, eight text and reference books. He has supervised more than 110 M.Sc. and 12 Ph.D. theses. His areas of interest are biometrics, image analysis, and processing and computer information systems. He has given 39 invited lectures and keynotes in different universities in Europe, China, India, South Korea, and Japan. The talks were on biometric image processing and analysis. He received about 18 academic awards. Khalid Saeed is a member of more than 15 editorial boards of international journals and conferences. He is an IEEE Senior Member and has been selected as IEEE Distinguished Speaker for 2011–2016. Khalid Saeed is the Editor-in-Chief of International Journal of Biometrics with Inderscience Publishers.

Nabendu Chaki is a Senior Member of IEEE and Professor in the Department of Computer Science and Engineering, University of Calcutta, India. Besides editing several volumes in Springer in LNCS and other series, Nabendu has authored three textbooks with reputed publishers like Taylor and Francis (CRC Press), Pearson Education, etc. Dr. Chaki has published more than 120 refereed research papers in journals and international conferences. His areas of research interests include image processing, distributed systems, and network security. Dr. Chaki has also served as a Research Assistant Professor in the Ph.D. program in Software Engineering at the Naval Postgraduate School, Monterey, CA, USA. He is a visiting faculty member for many universities including the University of Ca' Foscari, Venice, Italy. Dr. Chaki has contributed in SWEBOK v3 of the IEEE Computer Society as a Knowledge Area Editor for Mathematical Foundations. Besides being on the editorial board of Springer and many international journals, he has also served in the committees of more than 50 international conferences. He has been the founding Chapter Chair for the ACM Professional Chapter in Kolkata, India, since January 2014.

Part I
Security

Computer User Profiling Based on Keystroke Analysis

Tomasz Emanuel Wesołowski and Piotr Porwik

Abstract The article concerns the issues related to a computer user verification based on the analysis of a keyboard activity in a computer system. The research focuses on the analysis of a user's continuous work in a computer system, which constitutes a type of a free-text analysis. To ensure a high level of a users' data protection, an encryption of keystrokes was implemented. A new method of a computer user profiling based on encrypted keystrokes is introduced. Additionally, an attempt to an intrusion detection based on the k-NN classifier is performed.

Keywords Behavioral biometrics · Keystroke analysis · Free-text analysis · User verification

1 Introduction

The main task of the biometrics is the automatic recognition of individuals based on the knowledge of their physical or behavioral characteristics [1–4].

The behavioral biometrics methods use, among other things, an analysis of the movements of various manipulators (e.g., a computer mouse or a graphical tablet) [5] or the dynamics and statistics of typing on a computer keyboard [6–9]. The analysis of the way how a keyboard is used involves detection of a rhythm and habits of a computer user while typing on a keyboard [10]. The detection of these dependencies allows a recognition of a so-called user profile. This profile can then be used in the access authorization systems.

T.E. Wesołowski (✉) · P. Porwik
University of Silesia, Institute of Computer Science, Bedzinska 39,
41-200 Sosnowiec, Poland
e-mail: tomasz.wesolowski@us.edu.pl

P. Porwik
e-mail: piotr.porwik@us.edu.pl

© Springer India 2016
R. Chaki et al. (eds.), *Advanced Computing and Systems for Security*,
Advances in Intelligent Systems and Computing 395,
DOI 10.1007/978-81-322-2650-5_1

In the proposed method, the data stored in a user profile contains information on the sequence of keys and on time dependencies that occur between the key events. The advantage of the method is that collecting and analyzing a user's activity data is performed in the background, which makes it practically transparent for a user.

A computer user's profile can be used in a host-based intrusion detection system (HIDS). The task of a HIDS is to analyze the logs containing registered user's activity and the appropriate response when it detects an unauthorized access. These systems may analyze the profiled activity in a real time. A recognition of a user based on the analysis of the users' typing habits, while using a keyboard can effectively prevent an unauthorized access when a keyboard is overtaken by an intruder [11, 12].

This paper proposes a new method for creating a profile of the computer system user. It is assumed that the user works with a QWERTY type keyboard with standard layout of the keys. The encryption of the alphanumeric data entered via a keyboard is performed in order to prevent an access to users' passwords.

2   Reference Data Sets

2.1   Issues Related to Data Sets

There are a number of issues related to the data sets used in research on computer user profiling based on keystroke analysis. First of all, there are no standards for data collection and benchmarking as it is in some other fields of biometrics. For this reason, it is difficult to compare the works of different researchers.

The testing data sets used in experiments have some limitations. In most cases, the data sets are private and not available for other researchers [13, 14]. The form of a registered text differs between different approaches. Some researchers use in their study short phrases [15] or passwords [10, 16] that are typed many times by the same user while other register long texts [17, 18]. In case of long text analysis, users are asked either to copy a given text [18] or to type freely a limited number of characters [15, 19] using a software designed for this purpose. In the second case, it is a so-called "free-text" registration. However, a free-text recorded this way does not represent the situation when a user is typing while working with a computer on his tasks on a daily basis.

In order to develop an on-line type user profiling and intrusion detection method that could be implemented in a HIDS working in real time, it is necessary to analyze a continuous work of a computer user in an uncontrolled environment. There is some serious security issues connected to user's continuous activity registration. The approach presented in [13] is based on continuous work analysis. However, the data typed by the user is registered as an open text. As users often type private information (e.g., passwords, PINs) this approach constitutes a serious threat to a security of computer system. Another issue related to a continuous activity analysis is a registration software. So-called "key-loggers" are considered as malicious software. For this reason, a HIDS has to perform a data analysis on the fly (without storing the activity data) or if the activity data has to be stored encryption is necessary.

2.2 Data Acquisition

The biometric system presented in this paper is dedicated to and was tested in MS Windows operating systems. The identifiers of alphanumeric keys are encoded using the MD5 hash function. The same key identifier always receives the same code for the same user of the system.

The registration of user's activity data is performed automatically and continuously without involving a user. The data are captured on the fly and saved in the text files on the ongoing basis.

3 Data Analysis

The keyboard has been divided into groups of keys. The principle of the division is shown in Figs. 1 and 2.

The division of the keys is consistent with the following scheme for standard keyboard layout (Fig. 1):

- left function keys (with assigned identifiers $L1–L14$): $F1...F7$, *Esc, Tab, Caps lock, Left shift, Left ctrl, Windows, Left alt*;
- right function keys (with assigned identifiers $R1–R25$): $F8...F12$, *PrtScr, Scroll lock, Pause, Insert, Delete, Home, End, PgUp, PgDown, NumLck, Backspace, Enter, Right Shift, Right Ctrl, Context, Right alt, Arrows (up, down, left, right)*;
- alphanumeric keys (with assigned identifiers ID1–*ID*64);
- other keys.

In our approach, the tree structure includes 109 different groups G_{id}.

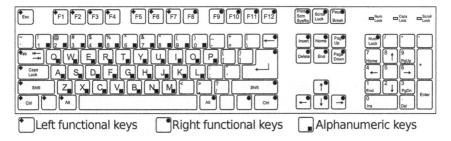

Fig. 1 QWERTY (105-keys) ISO standard keyboard with marked groups of keys

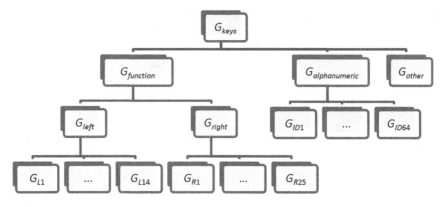

Fig. 2 Tree structure for organizing the groups of keys

Every use of the key is recorded in the next ith row of the text file as the following vector w_i:

$$w_i = [prefix, t_i, id] \tag{1}$$

where:

$prefix$—type of an event, $prefix \in \{\text{'}K\text{'}, \text{'}k\text{'}\}$, key down $\rightarrow$ 'K', key up $\rightarrow$ 'k',
t_i—time of an event,
id—key identifier (e.g., $ID1$, $L7$, $R15$, etc.).

4 User Profiling

4.1 Time Dependencies Extraction

The first stage of feature extraction is to convert the text data file containing a set of vectors w_i, into a form of time dependencies between the keyboard events. The data file is searched for the rows with identical identifier id, then each pair of rows containing one key down event and following one key up event is converted into a vector v_{id} according to the following principle:

$$\begin{cases} w_i = [\text{'}K\text{'}, t_i, id] \\ w_j = [\text{'}k\text{'}, t_j, id] \end{cases} \rightarrow v_{id} = [t_i, t_j], i < j. \tag{2}$$

Vectors w_i of the same type (with the same identifier id) should be present in the data file even number of times. Otherwise, the vector, for which the pair was not found, will be considered as an artifact and will be removed.

Vectors v_{id} containing the element id are assigned to the group G_{id} in a leaf of the tree structure presented in Fig. 2. After enrollment, the same vector v_{id} is added to all the groups higher in the branches of the tree structure until reaching the root group G_{keys}.

For example, if element id of a vector v_{id} is assigned an identifier $L1$ (it means that $id = L1$) than the mentioned vector v_{L1} will be added to the groups: G_{L1}, G_{left}, $G_{functional}$ and finally to G_{keys}.

4.2 Outliers Elimination

The user can use the keys of a keyboard freely, but the analysis of users activity is performed with some restrictions imposed on the key events. The next event cannot occur later than after the time t_{max} and the number of occurring consecutive events (that meet the first condition) cannot be less than c_{min}.

The values of parameters t_{max} and c_{min} have been determined experimentally.

For the exemplary keystroke sequence "$ABCDEF$" (Fig. 3) following times were recorded: $t_1 = 1.3$ s, $t_2 = 1.4$ s, $t_3 = 2.7$ s, $t_4 = 2.8$ s, $t_5 = 0.7$ s. Let $t_{max} = 2$ s and $c_{min} = 3$ be experimentally determined, than keystrokes 'A', 'B' and 'C' will be considered to be correct, and the associated vectors w_i will be added to the data set, from which in the future user's profile will be established. Other events will be considered as outliers and discarded (keystroke 'D' because $t_3 > t_{max}$ and keystrokes 'E' and 'F' because the number of elements in the group c_2 is lower than c_{min}).

4.3 Creating Feature Vectors

The next step of the activity data analysis is to create feature vectors. The groups G_{id} (Fig. 2) consist of vectors v_{id}. The total number of vectors that can be placed in the appropriate group is limited by g_{max}. The value of the g_{max} parameter has been determined experimentally and applies to all the groups of keystrokes in the tree structure.

When the number of vectors v_{id} assigned to the group G_{id} reaches its maximum value specified by g_{max}, then the feature vector is created. The group G_{id}, which recorded the number of vectors v_{id} equal to g_{max} is cleared and the process of adding further vectors v_{id} to the groups is resumed.

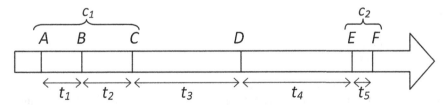

Fig. 3 The classification of keystrokes as outliers

The process ends when all the vectors v_{id} of a user have been processed or when the required number of feature vectors has been obtained.

4.4 Feature Vector

The feature vector is created based on the data contained in the groups G_{id} of the keystroke division structure (Fig. 2). For each group G_{id} standard deviation σ_{id} is calculated according to the formula (3).

Let the number of vectors $v_{id} = [t_i, t_j]$ registered in the group G_{id} be N. Then:

$$\sigma_{id} = \sqrt{\frac{1}{N}\sum_{k=1}^{N}(t_k - t_{id})^2} \tag{3}$$

where

t_k—dwell time of the k-th key belonging to the group G_{id} and $t_k = t_j - t_i$,
t_{id}—the average dwell time:

$$t_{id} = \frac{1}{N}\sum_{k=1}^{N}t_k \tag{4}$$

Finally, each feature vector consists of 109 standard deviation σ_{id} values (features). The above-described process is repeated and for a given user the next feature vector is created. The process is repeated as long as the required number of feature vectors has been obtained. The required number of feature vector was experimentally established. In our case, from the biometric point of view, the optimal number of feature vectors in a user's profile was equal to 1000.

5 The Results Obtained

The activity of four computer users has been registered within 1 month. In total, there are 123 files containing continuous activity data. The number of feature vectors generated for each user was

- *user*1—1470,
- *user*2—1655,
- *user*3—1798,
- *user*4—1057.

The feature vectors were normalized to the range of [0, 1]. Literature sources indicate a high efficiency of the k-NN classifiers [4, 6, 15, 16, 20, 21]. For this reason, intrusion detection was carried out by means of k-NN classifier.

Table 1 Parameter values used in the study

Parameter	Value
t_{max}	650 ms
c_{min}	5
g_{max}	15
k-NN	$k = 3$
Acceptance threshold	$\tau = 0.5$

The studies were verified using *leave-one-out* method and additionally repeated 20 times for different subsets of feature vectors of an intruder. The results obtained from the tests were averaged.

The testing procedure took into account the parameters described in Table 1. The values of parameters have been determined experimentally in order to obtain the lowest values of the EER.

5.1 Tuning the Parameters

In the first stage of the study, the experiments were performed to select the optimal values of the biometric system parameters. As an example, the results of experiments performed to obtain the optimal values of parameters k and t_{max} are presented. Figure 4 depicts the results of tuning the parameter k for the k-NN classifier.

Based on experiments the value of k was set to 3.

In Fig. 5, the results of experiments performed in order to obtain the optimal value of the t_{max} parameter are presented.

The best results were obtained for the value of the parameter t_{max} equal to 650 ms. For values below 400 ms, the outliers elimination process rejected too many keystrokes and there was not enough user's activity data left to create the necessary number of feature vectors for experiments.

5.2 The Final Results

The final results of the study are presented in Figs. 6 and 7. The charts in Figs. 6 and 7 should be interpreted as follows. The columns represent different pairs of computer owner (legitimate user) and intruder, for which the tests were performed. Each column indicates 20 intruder detection tests of the same intruder for different subsets of input data (dots). Squares represent the average score of 20 attempts for each pair of users. The dashed line represents the average value of the EER for the presented in this paper biometric system based on the analysis of the use of a computer keyboard.

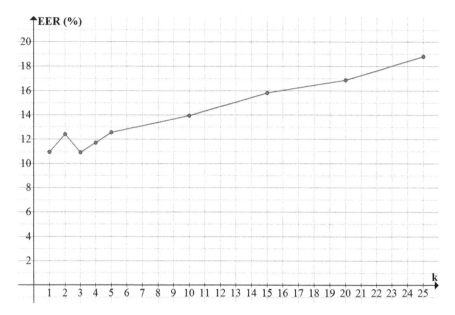

Fig. 4 The influence of parameter k on EER values of the method

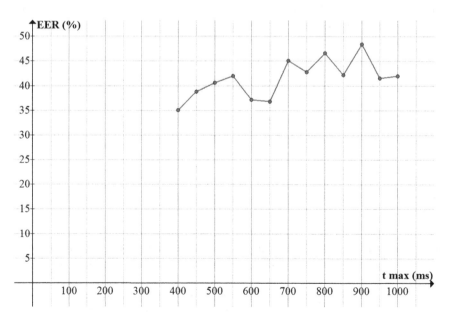

Fig. 5 The results of tuning the parameter t_{max}

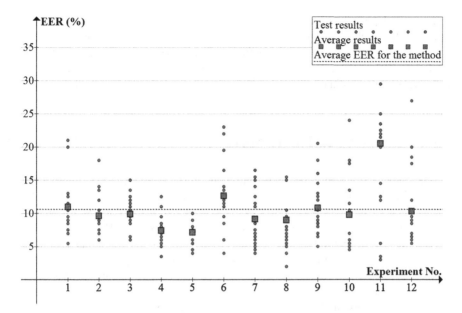

Fig. 6 Values of EER for the intrusion detection method and user's profile with outliers elimination

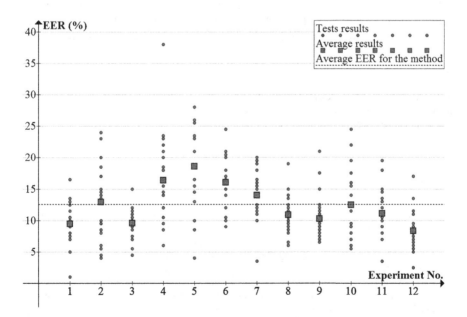

Fig. 7 Values of EER for the intrusion detection method and user's profile without outliers elimination

Figure 6 depicts the results of tests performed with the outliers elimination while in Fig. 7 the results of experiments conducted without eliminating the outliers are presented. The average value of the EER for all experiments was established at the level of 12.512 % without eliminating the outliers and at the level of 10.592 % with the outliers elimination.

6 Conclusions and Future Work

In this study, a series of experiments allowing an optimal selection of the parameters of the biometric system based on a use of a keyboard in order to determine the lowest value of the EER was performed. The achieved value of the EER equal to 10.59 % is better than the ones announced in [13, 16, 18, 19, 22].

The introduced method of a user's activity registration allows the analysis of a user's continuous work in an uncontrolled environment while performing everyday tasks. Additionally, a high security level was achieved by means of the MD5 hashing function.

Because the presented intrusion detection method uses a relatively high amount of data in order to create a user's profile and detect, the attack of a masquerader it is suitable for implementation in the off-line type intrusion detection systems.

In the future, the authors intend to explore the suitability of the other methods of data classification. As users often perform similar types of tasks during everyday activity future studies should consider the analysis of a user's activity in particular programs (e.g., text editors, web browsers). Additional research should be performed for users who work in the network environments where an intruder detection and localization is more difficult.

Acknowledgments The research described in this article has been partially supported from the funds of the project "DoktoRIS—Scholarship program for innovative Silesia" co-financed by the European Union under the European Social Fund.

References

1. Kudłacik, P., Porwik, P.: A new approach to signature recognition using the fuzzy method. Pattern Anal. Appl. **17**(3), 451–463 (2014). doi:10.1007/s10044-012-0283-9
2. Kudłacik, P., Porwik, P., Wesołowski, T.: Fuzzy Approach for Intrusion Detection Based on User's Commands. Soft Computing, Springer, Berlin Heidelberg (2015), doi:10.1007/s00500-015-1669-6
3. Pałys, M., Doroz, R., Porwik, P.: On-line signature recognition based on an analysis of dynamic feature. In: IEEE International Conference on Biometrics and Kansei Engineering, pp. 103–107, Tokyo Metropolitan University Akihabara (2013)
4. Porwik, P., Doroz, R., Orczyk, T.: The k-NN classifier and self-adaptive Hotelling data reduction technique in handwritten signatures recognition. Pattern Analysis and Applications, doi:10.1007/s10044-014-0419-1

5. Wesołowski, T., Pałys, M., Kudłacik, P.: computer user verification based on mouse activity analysis. Stud. Comput. Intell. **598**, 61–70 (2015). Springer International Publishing
6. Alsultan, A., Warwick, K.: Keystroke dynamics authentication: a survey of free-text methods. J. Comput. Sci. Issues **10**(1) 1–10 (2013) (Issue 4)
7. Araujo, L.C.F., Sucupira Jr., L.H.R., Lizarraga, M.G., Ling, L.L., Yabu-Uti, J.B.T.: User authentication through typing biometrics features. IEEE Trans. Signal Process. **53**(2) 851–855 (2005)
8. Banerjee, S.P., Woodard, D.L.: Biometric authentication and identification using keystroke dynamics: a survey. J. Pattern Recognit. Res. **7**, 116–139 (2012)
9. Teh, P.S., Teoh, A.B.J., Yue, S.: A survey of keystroke dynamics biometrics. Sci. World J. **2013**, Article ID: 408280, 24 pp. (2013) doi:10.1155/2013/408280
10. Zhong, Y., Deng, Y., Jain, A.K.: Keystroke dynamics for user authentication. In: IEEE Computer Society Conference, Computer Vision and Pattern Recognition Workshops, pp. 117–123 (2012), doi:10.1109/CVPRW.2012.6239225
11. Raiyn, J.: A survey of cyber attack detection strategies. Int. J. Secur. Its Appl. **8**(1), 247–256 (2014)
12. Salem, M.B., Hershkop, S., Stolfo, S.J.: A survey of insider attack detection research. Adv. Inf. Secur. **39**, 69–90, Springer US (2008)
13. Dowland, P.S., Singh, H., Furnell, S.M.: A preliminary investigation of user authentication using continuous keystroke analysis. In: The 8th Annual Working Conference on Information Security Management and Small Systems Security (2001)
14. Saha, J., Chaki, R.: An Approach to Classify Keystroke Patterns for Remote User Authentication. J. Med. Inf. Technol. **23**, 141–148 (2014)
15. Lopatka, M., Peetz, M.: Vibration sensitive keystroke analysis. In: Proceedings of the 18th Annual Belgian-Dutch Conference on Machine Learning, pp. 75–80 (2009)
16. Killourhy, K.S., Maxion, R.A.: Comparing anomaly-detection algorithms for keystroke dynamics. In: International Conference on Dependable Systems and Networks (DSN-09), pp. 125–134. IEEE Computer Society Press (2009)
17. Rybnik, M., Tabedzki, M., Adamski, M., Saeed, K.: An exploration of keystroke dynamics authentication using non-fixed text of various length, In: IEEE International Conference on Biometrics and Kansei Engineering, pp. 245–250 (2013)
18. Tappert, C.C., Villiani, M., Cha, S.: Keystroke biometric identification and authentication on long-text input. In: Wang, L., Geng, X. (eds.) Behavioral Biometrics for Human Identification: Intelligent Applications, pp. 342–367 (2010), doi:10.4018/978-1-60566-725-6.ch016
19. Gunetti, D., Picardi, C., Ruffo, G.: Keystroke analysis of different languages: a case study. Adv. Intell. Data Anal. **3646**, 133–144 (2005)
20. Foster, K.R., Koprowski, R., Skufca, J.D.: Machine learning, medical diagnosis, and biomedical engineering research—commentary. Biomed. Eng. Online **13**, 94 (2014), doi:10.1186/1475-925X-13-94
21. Hu, J., Gingrich, D., Sentosa, A.: A K-nearest Neighbor Approach for User Authentication through Biometric Keystroke Dynamics. In: IEEE International Conference on Communications, pp. 1556–1560 (2008)
22. Filho, J.R.M., Freire, E.O.: On the equalization of keystroke timing histogram. Pattern Recognit. Lett. **27**(13), 1440–1446 (2006)

Heart-Based Biometrics and Possible Use of Heart Rate Variability in Biometric Recognition Systems

Nazneen Akhter, Sumegh Tharewal, Vijay Kale, Ashish Bhalerao and K.V. Kale

Abstract Heart rate variability (HRV) is an intrinsic property of heart and active research domain of the medical research community since last two decades. But in biometrics it is still in its infancy. This article is intended to present the state of art into heart-based biometrics and also explore the possibility of using HRV in biometric recognition systems. Subsequently, we designed hardware and software for data collection and also developed software for HRV analysis in Matlab, which generates 101 HRV Parameters (Features) using various HRV analysis techniques like statistical, spectral, geometrical, etc., which are commonly used and recommended for HRV analysis. All these features have their relative significance in medical interpretations and analysis, but among these 101 features reliable features that can be useful for biometric recognition were unknown; therefore feature selection becomes a necessary step. We used five different wrapper algorithms for feature selection, and obtained 10 reliable features out of 101. Using the proposed 10 HRV features, we used KNN for classification of subjects. The classification test gave us encouraging results with 82.22 % recognition rate.

Keywords ECG biometrics · PCG biometrics · Heart rate variability · Linear HRV features · Poincare map · Feature selection · Wrapper algorithms · K-NN

N. Akhter (✉) · S. Tharewal · K.V. Kale
Department of Computer Science and Information Technology, Dr. Babasaheb Ambedkar Marathwada University, Aurangabad (MS), India
e-mail: getnazneen@gmail.com

S. Tharewal
e-mail: sumeghtharewal@gmail.com

K.V. Kale
e-mail: kvkale91@gmail.com

V. Kale · A. Bhalerao
M.G.M's G. Y. Pathrikar College of C.S & I.T, Aurangabad, India
e-mail: vijaykal1685@gmail.com

A. Bhalerao
e-mail: aashish.bhalerao@gmail.com

© Springer India 2016
R. Chaki et al. (eds.), *Advanced Computing and Systems for Security*,
Advances in Intelligent Systems and Computing 395,
DOI 10.1007/978-81-322-2650-5_2

1 Introduction

Biometric-based security systems are common nowadays and among all other modalities fingerprints, face, palm print, hand geometry, and voice by now are established and most utilized ones. Fundamental researches into these modalities have reached its pinnacle, yet certain propelled parameters are still being investigated to endeavor their utility to its fullest like in multidimensional space, multi- and hyperspectral space, etc. Similarly, even biosignals like electrocardiography (ECG), electroencephalogram (EEG), and electromyography (EMG), etc., are also being explored for improvising efficiency and robustness of biometric systems. For more than a decade, human heart is being explored as a potential candidate for biometric recognition. ECG being the most reliable clinical practice was the most explored for heart-based biometric purpose. Along with ECG, researchers have also focused on phonocardiogram (PCG) signals of heart (heart sounds) and recently even the heart pulses were studied using photoplethysmogram (PPG) signals.

Heart rate variability (HRV) is an intrinsic property of heart and active research domain of the medical research community since last two decades. But in biometrics it is still in its infancy. There have been several classification attempts for disease pattern identification in HRV data [1–4]. But only two early attempts of recognition using HRV are documented in the literature. Milliani et al. attempted to recognize two different postures, i.e., upright and supine of each individual using HRV [5], but mainly their focus was more on identification of posture and not specifically biometric recognition, while Irvine et al. proposed HRV-based human identification [6] which is the only reported attempt specifically aimed at biometric recognition but its techniques and results are unknown due to lack of information. This article is intended to present two independent parts, one is the state of art into heart-based biometrics and the second is to explore the possibility of using HRV in biometric recognition systems. Due to lack of HRV literature specifically dedicated to biometrics, we mostly had to rely on medical literature and using the techniques suggested by [7, 8]; we generated 101 HRV features (as per medical literature they are called as HRV parameters) which are commonly used and recommended for HRV analysis. All these features have their relative significance in medical interpretations and analysis, but among these 101 features, reliable features that can be useful for biometric recognition are unknown, therefore, feature selection becomes a necessary step, but it cannot be done arbitrarily hence we used five different wrapper algorithms for feature selection.

This article is divided into following sections: Sect. 2 will give a brief introduction to the background of HRV, while Sect. 3 will present state of art into heart-based biometrics, further Sect. 4 will present a much elaborated methodology section including the hardware and software designing as well as an extensive feature sets generation process. And finally, the last sections include the results and discussions along with conclusion and future directions.

2 HRV Background

The time interval (duration/gap) between two adjacent R–R peaks of QRS complex of a heartbeat is known as R–R interval as shown in Fig. 1. This R–R interval varies in every adjacent pair of beat as shown in Fig. 2. The variance in RRI, i.e., beat-to-beat variation is popularly known as heart rate variability (HRV). HRV analysis is the ability to assess the overall cardiac health and the state of the autonomic nervous system (ANS) responsible for regulating cardiac activity [8]. HRV is a useful signal for understanding the status of the ANS. The balancing action of the sympathetic nervous system (SNS) and parasympathetic nervous system (PNS) branches of the ANS controls the HR. Increased SNS or diminished PNS activity results in cardioacceleration. Conversely, a low SNS activity or a high PNS activity causes cardio-deceleration. The past decades have witnessed the

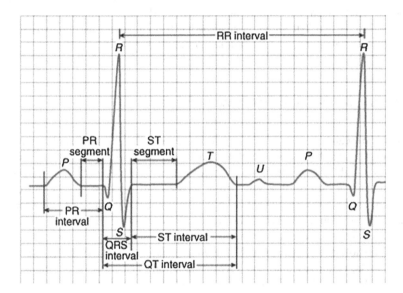

Fig. 1 Schematic of an ECG *strip* showing major components

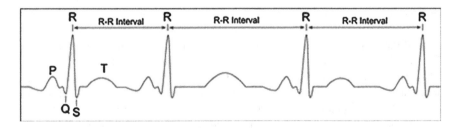

Fig. 2 Schematic of ECG indicating the R–R interval variation

recognition of the significant relationship between ANS (automatic nervous system) and cardiovascular mortality, including sudden death due to cardiac arrest [7]. In 1996, a task force of the European Society of Cardiology and the North American Society of Pacing and Electrophysiology developed and published standards for the measurement, physiological interpretation, and clinical use of HRV analysis [8].

Heart rate (HR) or the beat interval is a nonstationary signal; its variation may contain indicators of current disease, or warnings about impending cardiac diseases. The indicators may be present at all times or may occur at random during certain intervals of the day. It is strenuous and time consuming to study and pinpoint abnormalities in volumes data collected over several hours. Therefore, the HRV signal parameters, extracted and analyzed using computers, are highly useful in diagnostics. Computer-based analytical tools for in-depth study of data over day-long intervals can be very useful in diagnosis. Analysis of HRV consists of a series of measurements of successive R–R interval variations of sinus origin, which provide information about autonomic tone [9]. Different physiological factors may influence HRV such as gender, age, circadian rhythm, respiration, and body position [10]. Hence, HR variation analysis has become a popular noninvasive tool for assessing the activities of the autonomic nervous system.

3 State-of-the-Art Heart-Based Biometrics

Biometric research community is exploring every physiological and behavioral aspect of humans that can be employed in biometric recognition systems. Human heart has caught the attention since more than a decade. Heart-based biometric research activities up till now are mainly focused on two very specific aspects of heart, one is the sound produced by heart, i.e., lub, dub sound and the electrical signal generate from every beat of heart, i.e., electrocardiogram (ECG) and recently even PPG signals were also examined. Heart sounds are one of the most important human physiological signals, which contain information about the atrium, ventricle, great vessels, cardiovascular, and valvular function [11]. Most of the initial literature available on heart sound-based biometric recognition is contribution of Beritelli et al. [11–15]. In 2007 Beritelli et al. [12] first time proposed the frequency analysis of heart sounds for the identity purpose. And in 2009 Beritelli et al. applied MFCC technique on heart sounds [13] and reported a 10 % reduction in the EER rate from their previous work. By the year 2010, few more researchers Huy Dat et al. [16], Ye-wei [17], Al-Shamma et al. [18], Jasper [19] and even Beritelli [12] extended the work. Tao et al. attempted to extract features using wavelet and used CPSD for recognition purpose, while Huy Dt. et al. attempted fusion of features. Al-Shamma et al. used energy percent in wavelet coefficients. And Beritelli et al. applied statistical approaches. Then in 2011, Zhidong Zhao and Jia Wang proposed MFCC with vector quantization. In 2013 Gautam and Kumar [20], proposed feature set based on Daubechies wavelet with second-level decomposition and did classification using Back Propagation Multilayer Perceptron Artificial Neural Network

(BP-MLP-ANN) classifier. And recently in 2014, Abo-Zahhad et al. [21] also attempted feature fusion, but applied canonical correlation analysis.

The pioneering work on ECG-based biometrics recognition is credited to Biel et al. [22] who in 2001 not only proposed that the ECG of a person contains sufficiently detailed information which is highly personalized, but also that single channel ECG is sufficient for biometric purpose. Use of single channel ECG is the simplest hence most studied. However, some researchers have documented improved results by incorporating 2, 3, and even 12 channels [23–26]. Traditional clinical grade ECG devices, though may be too advance, but are too complex for biometrics systems with poor user acceptability, there have been a dart of specifically designed standard ECG database for biometrics research; so far what researchers have been using is either physionet [27] a central repository having huge collection of ECG records of healthy subjects and even with pathological conditions, or MIT-BIH [28] which has also served the research community in their ECG-based research endeavors but all records coming from clinical settings. Both these databases are not designed specifically keeping biometrics in mind. Most of researchers face a common problem of lack of larger databases to test their hypothesis. And this is the special case with ECG-based biometrics is being studied. Hugo et al. in [29] made some efforts in resolving the database shortage issue hovering over the ECG-based biometric research by creating the CYBHi ECG database particularly for biometric research. Also Lourenco et al. in [30] proposed that ECG collected at finger tips are sufficiently enough for biometric recognition. Hugo Silva et al. in [31] presented a very simple approach toward ECG biometrics by subjecting the ECG strips to segmentation and creating mean and median waves and using them as templates in authentication systems.

R–R intervals are the duration between two consecutive heart beats as in Fig. 1, this duration represents the variability property of heart rate. Only R–R interval is required for HRV analysis and it is traditionally measured from ECG signals. Researchers have documented evidences in favor of PPG to surrogate ECG for HRV analysis [32–34]. While the ECG monitors the electrical activity, PPG monitors the mechanical activity of the same event, i.e., a heartbeat. ECG from the chest is the clearest, but rarely used outside hospital [35] and if it has to be employed in biometric applications it faces the challenge of poor user cooperation. If heart signals are to be used in biometric recognition systems, then other methods need to be explored. PPG sensors being low cost and comfortable in data collection are one of the instant choices for ECG alternative. While Gu et al. [36] proposed a novel biometric approach for human verification using PPG signals, Resit et al. [37] proposed a novel feature ranking algorithm for time domain features from first- and second-order derivatives of PPG signals for biometric recognition.

Israel et al. in [38] gave an extensive performance analysis of three different sensing methods of heart, i.e., ECG, pulse oximetry, and blood pressure, which documented the latter two methods to be on the lower side. da Silva et al. in [39] has presented the usability and performance study of heart signals from fingertips and also in [40] da Silva et al. proposed CYBHi (Check Your Biosignals Here initiative) ECG dataset which was collected at fingertips of the subjects.

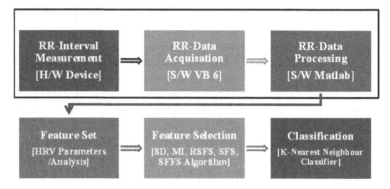

Fig. 3 Workflow of proposed HRV biometric system

4 Proposed System

In the proposed system, as seen in Fig. 3, initially the biometric data of subjects is collected in this case it is the raw HRV time series data, i.e., the R–R interval sequence. For data acquisition, we designed and developed a R–R interval measurement hardware that is equipped with a pulse sensor to detect the heartbeats. This gave us the freedom to generate our own KVKHRV database, as there is no such specific standard HRV database available for biometric purpose.

4.1 Block Diagram

The first step in the process is the detection of the R–R interval. The detector hardware based on a light sources and a detector employing the signal conditioning electronics senses the heartbeat and produce pulses in synchronism with the heartbeats. The interval between two consecutive beats, i.e., the R–R interval is implemented using a microcontroller-based circuit that measures the time interval between two consecutive beats (R–R interval) in milliseconds and sends it to serial port using RS232 protocol that can be received by any standard device like a computer. This R–R interval is received by a computer to which the hardware is interfaced, the computer side controlling program is developed in visual basic. This data acquisition software (shown is second block) unscrambles the incoming data and performs the necessary processing and saves it in standard text files for further use and processing. In the next block, the files from data acquisition system are further processed in programs specifically developed in Matlab12 for this purpose. This software allows for selecting parameters required for implementation of different algorithms like statistical, spectral, time-frequency, and nonlinear techniques, for the purpose of extraction of features for use in the authentication system. In the next step, feature selection takes place using five different algorithms and finally the last block, i.e., classification is done using KNN classifier.

4.2 R–R Interval Acquisition

Basically, the detection hardware detects the heart beats and the associated microcontroller-based system through its firmware polls for the arrival of a pulse and computes the time in milliseconds elapsed between the arrivals of two consecutive pulses. This R–R interval is sent through the serial port to a computer interfaced to the acquisition system via a serial to USB bridge. The computer side controlling program receives this RR data via USB port in the form of two bytes and performs some preprocessing like combining two bytes and saves the results in text files for further use.

The data acquisition system of Fig. 4 is the computer side program written in visual basic 6 that collects the R–R Intervals and stores in text files for further processing. This is developed in Visual Basic with GUI support for ease of operation in a user friendly manner and displays the real time R–R interval received in a graphic panel. Screenshot in Fig. 4 shows a typical data collection for 512 intervals of a subject. The GUI consists of four modes of collecting R–R intervals, i.e., for 1, 2, 5, and 10 min; in the first mode, 64 R–R intervals are measured for 128 intervals, similarly 128, 256, and 512, respectively. The computer side program

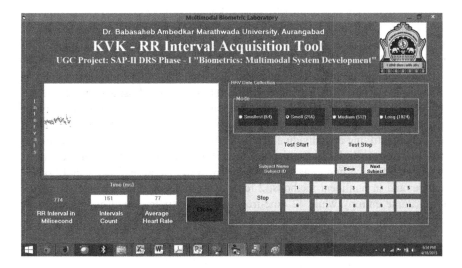

Fig. 4 GUI for R–R interval acquisition system

has provision for recording data from different subjects recorded under different conditions in the appropriate folders for further analysis. It also has the capability to auto detect and remove ectopic beats, i.e., noise from the HRV data.

4.3 Database Specification

At present, our in house-generated database consists of sequence of R–R intervals of 81 subjects (47 males and 34 Females) whose 512 R–R intervals were measured continuously for 10–12 min approximately in session one, while 64 R–R intervals were measured continuously for 1–1.5 min approximately in sessions two and three with time interval of 3 months between each session. The age of the individuals varied from 18 to 69 years, with mean and standard deviation of 31 and 11, respectively. As it would be natural in any physiological-based biometric recognition system, some subjects would have health issues; we too have few samples of this sort around 11 % of subjects reported hypertension and other diseased conditions. Any biosignals-based biometric system is susceptible to the effects of mental, physical, physiological, and even emotional state of the subject. Hence, subjects were relaxed first and data was collected in sitting relaxed position for all the sessions.

4.4 Feature Set Generation

Experimenting HRV for biometric recognition we generated the HRV parameters suggested in [4] with a few more additions identified from the literature survey. HRV parameters are actually the results of applying various linear methods like statistical and spectral techniques and nonlinear like Poincare and auto regression and also some time-frequency methods like wavelets on RR data. These HRV parameters can serve as a feature vector for biometric classification. In all we obtained 101 features, each has some significance or the other in HRV analysis for diagnostic or prognostic purpose, but which one would really prove suitable for uniquely identifying an individual that is yet to be established. The initial feature set of 101 features includes 9 statistical features 39 frequency domain features obtained by applying three different techniques, namely Welch, auto regression, and Lomb–Scargle; and in nonlinear methods, 2 features from a Poincare map and 4 features from sample entropy while 42 time-frequency analysis features from Welch, auto regression and wavelet power spectrum density analysis (Fig. 5).

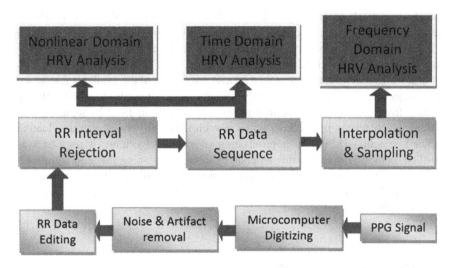

Fig. 5 R–R interval processing for HRV features set generation

4.5 Feature Selection Algorithms

Biometric applications inherently are pattern-recognition problems. And in any pattern recognition system, each pattern is represented by a set of features or measurements and is viewed as a point in the dimensional feature space [19] called as feature vectors. As HRV-based biometrics recognition is being explored for the first time we are not sure whether all or subsets of features will give best classification. Apparently, feature selection becomes a very critical and major step before the classification. Actually, as a matter of fact feature selection is an important problem for pattern classification systems [20], which aims at selecting features that allow us to discriminate between patterns belonging to different classes and in biometrics it aims at discriminating or recognizing different humans. The feature set with 101 HRV features generated by HRV analysis naturally must contain irrelevant or redundant features which would degrade the performance of classification. In general, it is desirable to keep the number of features as small as possible to reduce the computational cost of training a classifier as well as its complexity [19] in addition of getting a better classification rate. According to Jain et al. [21], feature extraction methods create new features based on transformations or combinations of the original feature set, whereas feature selection refers to methods that select the best subset of the original feature set. Feature selection algorithms can be classified into filters and wrappers [22]. Filter methods select subset of features as a preprocessing step, independent of the induction (learning) algorithm. Wrappers utilize the classifier (learning machine) performance to evaluate the goodness of feature subsets [19]. As Wrapper methods are widely recognized as superior alternative in supervised learning problems [23], we choose five wrapper methods, namely statistical dependency (SD) which estimates the statistical dependency between features and associated class labels using a quantized feature space [19, 23], mutual information (MI) measures arbitrary dependencies between random variables [19, 24], random subset feature selection (RSFS) aims to discover a set of features that perform better than an average feature of the available feature set [19], sequential forward selection (SFS) works in the opposite direction, initially starting from an empty set, the feature set is iteratively updated by including the feature which results in maximal score in each step [19], and sequential floating forward selection (SFFS) is improvisation over SFS algorithm, it uses SFS as baseline method [19] and further extends by iteratively finding the least significant features and eliminates it; this process continues till a desired number of features are not obtained. The obtained features are discussed in detail in the results and discussion section.

5   Results and Discussions

From the raw data files of R–R interval sequence, feature set generation was implemented using statistical, spectral, time-frequency, and nonlinear techniques like Poincare map and sample entropy. In all 101 features were generated out of which 14 from statistical techniques, 39 from three different spectral techniques, and 42 from three different time-frequency techniques and remaining from nonlinear techniques.

It was observed that not all the features from all the techniques are very much relevant and these features also depend on the nature of the raw data used. Many features are proposed from different considerations and point of view, and for different applications which may not prove to be effective in the present context of biometric recognition. All features described above are obtained from HRV analysis used in diagnostics, and therefore the main concern is which of these features are going to be effective in biometric recognition. Extensive efforts have been put in for arriving at a rationale to select features that have relative significance and are promising in biometric recognition. With this in view, we subjected the complete feature set to five different tests based on SD, MI, RSFS, SFS, and SFFS algorithms.

From the entire set of features, the first two tests suggested a list of 10 best features while the third one gave 23 and the fourth and fifth one gave 10 sensitive features each. The fives lists of selected features suggested by above algorithms partly overlapped as seen in Table 1. We categorized the features appearing in all the five groups described above as strong, those appearing in three to four groups were considered as moderately well, those found in two groups as weak and features suggested by only one test were considered as poor and were set aside. Features appearing in all 5 and 3, 4 groups are listed in the Table 2. It was found that the range of values covered by the features is large enough, some of the features have values in fractions whereas others are in thousands. This broad range of values resulted in poorer comparison that was evident from the performance in

Table 1 List of features suggested by five wrapper algorithms (for names and descriptions of features, please see Appendix A)

S. no.	Algorithm	No. of features in proposed features list	Features proposed in features list of wrapper algorithms
1	Statistical dependency (SD)	10	max, mean, median, RMSSD, meanHR, aHF (welch), aHF(Burg), SD1, aHF (lomb), aHF(wavelet)
2	Mutual information (MI)	10	max, min, mean, median, RMSSD, meanHR, aHF (welch), SD1, aHF (lomb), aHF(wavelet)
3	Random subset feature selection (RSFS)	23	max, min, mean, median, SDNN, RMSSD, meanHR, sdHR, aTotal Welch), aHF(Burg), aTotal (Burg), peakHF(lomb-Scargle), SD1, SD2, aHF (Burg), aTotal (Burg), peakHF (Burg), aHF (lomb), aTotal (lomb), peakHF (lomb), aLF (Wavelet), aHF(wavelet), aTotal (Wavelet)
4	Sequential forward selection (SFS)	11	mean, RMSSD, meanHR, sdHR, HRVTi, SD1, SD2, aHF (Burg), pHF (Burg), LFHF (lomb), LFHF (Wavelet)
5	Sequential floating forward selection (SFFS)	11	mean, RMSSD, meanHR, sdHR, HRVTi, SD1, SD2, aHF (Burg), pHF (Burg), LFHF (lomb), LFHF (Wavelet)

Table 2 List of features appearing in 3 or more groups

S. no.	Feature(s) name	Technique name
1	Max, mean, median, RMSSD, meanHR, sdHR	Statistical technique
2	SD1, SD2	Poincare chart
3	aHF	Spectral (Lomb)
4	aHF	Time-frequency (Wavelet)

the classification test. This suggested a comparison of features on a similar scale by way of normalization, when normalized the features gave a better performance in the classification test. Details are shown in Table 3.

Using the criterion discussed above, selecting ten features (occurrence 3–5 times in the suggested feature list) KNN classifier was tested on the feature set of 27 subjects. It was found that the results significantly improved to 82.22 % of the testing set.

Table 3 List of features appearing in 3 or more groups

S. no.	Algorithm	Recognition rate (%)	
		Without normalization	With normalization
1	SD	60.37	66.30
2	MI	59.26	68.89
3	RSFS	49.63	65.93
4	SFS	63.70	68.89
5	SFFS	63.70	64.63

6 Conclusions and Future Directions

Heart being a vital organ containing characteristic properties for each individual proves to be a potential candidate for biometric recognition. Different approaches have been proposed utilizing different properties like its sound and its electrical activity. One of the important characteristic of the heart is its heart rate variability (HRV) that has been used for different applications including diagnosis and prognosis. We attempted feature generation using different techniques like statistical, spectral, time-frequency, and nonlinear like Poincare and sample entropy used in HRV analysis. In all 101 features have been obtained and to pinpoint the features that are promising in terms of biometric identification we used SD, MI, RSFS, SFS, and SFFS feature selection techniques. After identification of the features, ten prominent features were selected that were common to more than two selection algorithms.

Initial work showed that the range of values of different features extracted is very large, there are features with fractional values, whereas others are in thousands. This suggested that the features are to be compared on similar scale, for this the features were normalized and the normalization resulted in improved results as shown in Table 2. The recognition rate with the ten features found in more than two groups using KNN classifier gave 82.22 % for the testing set.

Looking at the performance of the selected features, it appears that HRV-based biometric recognition is promising research area which needs more prospective studies with larger databases and context aware data conditions. Performance of KNN is seen in the present work, but more classifiers can also be experimented to improve the results further. HRV data can also be used in liveness detection hence attempts in those directions would yield interesting results. HRV can also be experimented in multimodal system and is expected to add much needed robustness and efficiency. With little modification in hardware and data acquisition software, the same setup can also be used for continuous authentication. Due to a simple user friendly device we designed, all these research dimensions look achievable.

Acknowledgments This work was carried out in Multimodal System Development laboratory established under UGC's SAP scheme, SAP (II) DRS Phase-I F. No. 3-42/2009 & SAP (II) DRS Phase-II F. No.4-15/2015. This work was also supported by UGC under One Time Research Grant F. No. 4-10/2010 (BSR) & 19-132/2014 (BSR). The authors also acknowledge UGC for providing BSR fellowship.

Appendix A

Poincare map features		
1	• SD1	• Standard deviations of the distances of the R–R (I) to the lines
2	• SD2	• Y = x and y = –x + 2R–Rm, where R–Rm is the mean of all R–R (I)
		• SD1 related to the
		• Fast beat-to-beat variability in the data, while SD2
		• Describes the long-term variability of R–R (I)

Statistical features		
No.	Name	Description
1	SDNN	Standard deviation of all normal–normal intervals
2	RMSSD	Root mean square of successive differences
3	NN50	It's a count of the number of adjacent pairs differing by more than 50 ms
4	pNN50	(%) NN50 count divided by total intervals
5	MeanRRI	Mean of normal–normal interval
6	MeanHR	Mean heart rate
7	Max	Maximum interval duration in a particular RRI
8	Min	Minimum interval duration
9	Mean	Mean of the whole RRI sequence
10	Median	Median of the RRI sequence
11	SDHR	Standard deviation of heart rate

Spectral features		
No.	Name	Description
1	aVLF	Absolute value in very low-frequency spectrum
2	aLF	Absolute value in low-frequency Spectrum
3	aHF	Absolute value in high-frequency Spectrum
4	aTotal	Total absolute value
5	pVLF	Power % of very low frequency in PSD
6	pLF	Power % of low frequency in PSD
7	pHF	Power % of high Frequency in PSD
8	nLF	Low frequency in normalized Unit
9	nHF	High frequency in normalized Unit
10	LFHF	LF to HF Ratio
11	peakVLF	Peak value in very low frequency
12	peakLF	Peak value in low frequency
13	peakHF	Peak value in high frequency

References

1. Lin, C., Wang, J.-S., Chung, P.: Mining physiological conditions from heart rate variability analysis (2010)
2. Melillo, P.: Classification tree for risk assessment in patients suffering from congestive heart failure via long-term heart rate variability. IEEE J. Biomed. Heal. Inform. **17**, 727–733 (2013)
3. Nizami, S., Green, J.R., Eklund, J.M., McGregor, C.: Heart disease classification through HRV analysis using parallel cascade identification and fast orthogonal search. In: Proceedings of 2010 IEEE International Workshop on Medical Measurements and Applications, MeMeA 2010, pp. 134–139 (2010)
4. Szypulska, M., Piotrowski, Z.: Prediction of fatigue and sleep onset using HRV analysis. In: Proceedings of the 19th International Conference Mixed Design of Integrated Circuits and Systems (MIXDES), pp. 543–546 (2012)
5. Malliani, A, Pagani, M., Furlan, R., Guzzetti, S., Lucini, D., Montano, N., Cerutti, S., Mela, G. S.: Individual recognition by heart rate variability of two different autonomic profiles related to posture. Circulation **96**, 4143–4145 (1997)
6. Irvine, J.M., Wiederhold, B.K., Gavshon, L.W., Israel, S.A., McGehee, S.B., Meyer, R., Wiederhold, M.D.: Heart rate variability: a new biometric for human identification. In: Proceedings of the International Conference on Artificial Intelligence IC-AI'2001, pp. 1106–1111(2001)
7. AHA and ESC: Guidelines heart rate variability. Eur. Heart J. 354–381 (1996)
8. Acharya, U.R., Joseph, K.P., Kannathal, N., Lim, C.M., Suri, J.S.: Heart rate variability: a review. Med. Biol. Eng. Comput. **44**, 1031–1051 (2006)
9. Chang, F.C., Chang, C.K., Chiu, C.C., Hsu, S.F., Lin, Y.D.: Variations of HRV analysis in different approaches (2007)
10. Aletti, F., Ferrario, M., Almas de Jesus, T.B., Stirbulov, R., Borghi Silva, A., Cerutti, S., Malosa Sampaio, L.: Heart rate variability in children with cyanotic and acyanotic congenital heart disease: Analysis by spectral and non linear indices (2012)
11. Spadaccini, A., Beritelli, F.: Performance evaluation of heart sounds biometric systems on an open dataset (2013)
12. Beritelli, F., Serrano, S.: Biometric identification based on frequency analysis of cardiac sounds (2007)
13. Beritelli, F., Serrano, S.: Biometric identification based on frequency analysis of cardiac sounds. IEEE Trans. Inf. Forensics Secur. **2**, 596–604 (2007)
14. Beritelli, F., Spadaccini, A.: Heart sounds quality analysis for automatic cardiac biometry applications. Francesco Beritelli and Andrea Spadaccini Dipartimento DI Ingegneria Informatica e delle Telecomunicazioni, University of Catania, Italy, pp. 61–65 (2009)
15. Beritelli, F., Spadaccini, A.: An improved biometric identification system based on heart sounds and Gaussian mixture models (2010)
16. Tran, H.D., Leng, Y.R., Li, H.: Feature integration for heart sound biometrics. In: 2010 IEEE International Conference on Acoustics Speech and Signal Processing ICASSP, pp. 1714–1717 (2010)
17. Ye-wei, T.Y.T., Xia, S.X.S., Hui-xiang, Z.H.Z., Wei, W.W.W.: A biometric identification system based on heart sound signal. In: 2010 3rd Conference on Human System Interaction (HSI), (HSI), pp. 67–75 (2010)
18. Al-Shamma, S.D., Al-Noaemi, M.C.: Heart sound as a physiological biometric signature. In: 2010 5th Cairo International Biomedical Engineering Conference, pp. 232–235 (2010)
19. Jasper, J., Othman, K.R.: Feature extraction for human identification based on envelogram signal analysis of cardiac sounds in time-frequency domain (2010)
20. Gautam, G.: Biometric System from heart sound using wavelet based feature set, pp. 551–555 (2013)

21. Ahmed, S.M., Abbas, S.N., Engineering, E.: PCG Biometric identification system based on feature level fusion using canonical correlation analysis, pp. 1–6 (2014)
22. Biel, L., Pettersson, O., Philipson, L., Wide, P.: ECG analysis: a new approach in human identification (2001)
23. Wübbeler, G., Stavridis, M., Kreiseler, D., Bousseljot, R.-D., Elster, C.: Verification of humans using the electrocardiogram. Pattern Recognit. Lett. **28**, 1172–1175 (2007)
24. Agrafioti, F., Hatzinakos, D.: Fusion of ECG sources for human identification (2008)
25. Ye, C., Coimbra, M.T., Kumar, B.V.K.V.: Investigation of human identification using two-lead electrocardiogram (ECG) signals (2010)
26. Fang, S.-C., Chan, H.-L.: Human identification by quantifying similarity and dissimilarity in electrocardiogram phase space. Pattern Recognit. **42**, 1824–1831 (2009)
27. Oeff, M., Koch, H., Bousseljot, R., Kreiseler,D.: The PTB Diagnostic ECG Database, National Metrology Institute of Germany. http://www.physionet.org/physiobank/database/ptbdb/. Accessed 19 June 2015
28. The MIT-BIH Normal Sinus Rhythm Database, http://www.physionet.org/physiobank/database/nsrdb/. Accessed 19 June 2015
29. da Silva, H.P., Lourenço, A., Fred, A., Raposo, N., Aires-de-Sousa, M.: Check your biosignals here: a new dataset for off-the-person ECG biometrics. Comput. Methods Programs Biomed. **113**, 503–514 (2014)
30. Lourenço, A., Silva, H., Santos, D.P., Fred. A.L.N.: Towards a finger based ECG biometric system. Biosignals 348–353 (2011)
31. da Silva, H.P., Lourenço, A., Canento, F., Fred, A., Raposo, N.: ECG Biometrics: principles and applications. In: Proceedings of International Conference on Bio-inspired Systems and Signal Processing—Biosignals—INSTICC (2013)
32. Lin, W.-H., Wu, D., Li, C., Zhang, H., Zhang, Y.-T.: Comparison of Heart Rate Variability from PPG with That from ECG. In: Zhang, Y.-T. (ed.) The International Conference on Health Informatics SE—54, pp. 213–215. Springer International Publishing (2014)
33. Selvaraj, N., Jaryal, A., Santhosh, J., Deepak, K.K., Anand, S.: Assessment of heart rate variability derived from finger-tip photoplethysmography as compared to electrocardiography. J. Med. Eng. Technol. **32**, 479–484 (2008)
34. Gil, E., Orini, M., Bailón, R., Vergara, J., Mainardi, L., Laguna, P.: Photoplethysmography pulse rate variability as a surrogate measurement of heart rate variability during non-stationary conditions, Physiol. Meas. **31**(9), 127–1290 (2010)
35. Park, B.: Psychophysiology as a tool for HCI Research: promises and pitfalls. Human-Computer Interaction. New Trends SE—16, vol. 5610, pp. 141–148 (2009)
36. Gu, Y.Y., Zhang, Y., Zhang, Y.T.: A novel biometric approach in human verification by photoplethysmographic signals. In: 4th International IEEE EMBS Special Topic Conference on Information Technology Applications in Biomedicine, pp. 13,14 (2003)
37. Reşit Kavsaoğlu, A., Polat, K., Recep Bozkurt, M.: A novel feature ranking algorithm for biometric recognition with PPG signals. Comput. Biol. Med. **49**, 1–14 (2014)
38. Israel, S.A., Irvine, J.M., Wiederhold, B.K., Wiederhold, M.D.: The heartbeat: the living biometrics. Theory, Methods, Appl. 429–459 (2009)
39. da Silva, H.P., Fred, A., Lourenco, A., Jain, A.K.: Finger ECG signal for user authentication: usability and performance. Biometrics: Theory, Appl. Syst. (2013)
40. da Silva, H.P., Lourenço, A., Fred, A., Raposo, N., Aires-de-Sousa, M.: Check your biosignals here: a new dataset for off-the-person ECG biometrics. Comput. Methods Programs Biomed. **113**, 2503–514 (2014)

Dynamic Ciphering-15 Based on Multiplicative Polynomial Inverses Over Galois Field GF(7^3)

J.K.M. Sadique Uz Zaman, Sankhanil Dey and Ranjan Ghosh

Abstract A new stream ciphering technique based on multiplicative polynomial inverses over Galois Field GF(7^3) is proposed, where a set of randomly generated key-bytes, between 1 and 15, is dynamically permuted and XORed with the identical number of message bytes. The output cipher is tested using NIST Statistical Test Suite and results are compared with that obtained by the well-known RC4 stream cipher. The new cipher is statistically random and observed to be better than RC4.

Keywords Dynamic ciphering · Extension field · Galois field · GF(7^3) · Multiplicative polynomial inverse · NIST statistical test · Randomness · RC4

1 Introduction

In this paper, multiplicative polynomial inverses under an irreducible polynomial over Galois Fields GF(7^3), available in [1], are innovatively used to design a new dynamic stream cipher following the technique of randomly shuffling S-Box elements adopted in RC4 [2, 3]. The initial identity S-Box of RC4 is replaced by a nonidentity S-Box obtained from multiplicative polynomial inverses over GF(7^3). An additional process is incorporated where a set of generated random key-bytes between 1 and 15 is dynamically permuted among themselves. Following the permutation process, a

J.K.M.S.U. Zaman (✉) · S. Dey · R. Ghosh
Department of Radio Physics and Electronics, University of Calcutta,
92 A. P. C. Road, Kolkata 700 009, India
e-mail: jkmsadique@gmail.com

S. Dey
e-mail: sankhanil12009@gmail.com

R. Ghosh
e-mail: rghosh47@yahoo.co.in

© Springer India 2016　　　　　　　　　　　　　　　　　　　　　　31
R. Chaki et al. (eds.), *Advanced Computing and Systems for Security*,
Advances in Intelligent Systems and Computing 395,
DOI 10.1007/978-81-322-2650-5_3

group XORing operation is executed on the set of permuted key-bytes with identical number of message bytes replacing the character-by-character XORing operation of RC4. The output cipher is tested using 15 randomness tests proposed in the NIST statistical tests suite and results are compared with that obtained using RC4. It is observed that the new cipher is also statistically random and quantitatively better than that obtained with RC4.

RC4 algorithm is simple in which all additions are 256 modulo additions. It starts from an 8-bit identity S-Box and the given key elements are repetitively stored in an 8-bit K-Box. The RC4 undertakes random shuffling of the S-Box elements first using key elements stored in the K-Box and then without the key elements in order to systematically increase the arrangement of S-Box elements more and more random. A sequential index (i) and a random index (j) are defined. RC4 assumes both the indices as zero and enters a KSA loop being executed 256 times. In each loop, j is upgraded by addition of itself with $S[i]$ and $K[i]$, followed by swapping of $S[i]$ and $S[j]$. After KSA, RC4 enters an infinite PRGA loop in which both i and j starts from zero and in each PRGA loop i is upgraded by adding unity and j is upgraded by addition of itself with $S[i]$ only followed by swapping of $S[i]$ and $S[j]$—the result of addition of $S[i]$ and $S[j]$ is used as an index of random key-byte. Many researchers [4–8] observed various types of weak keys. Even for good keys, they also observed key bias [4, 6] in few initial PRGA bytes and suggested many modifications in RC4 [7, 8] in order to overcome the weakness. They also suggested that the conventional RC4, with no modifications whatsoever, would exhibit better performance without key bias if few initial PRGA bytes are discarded, possibly 256 as suggested by Roos [4]; but according to Preneel [6], the said amount should be at least 512 while it is 1024 as per Maitra [8]. Following a precise look, one would be convinced to notice two loopholes in the RC4 algorithm behind the weakness of key bias: (1) considering initial S-Box with identity elements and (2) repetitive insertion of given key elements all through the K-Box.

One can overcome the weakness of key bias, if the initial identity S-Box is replaced by a nonidentity S-Box and the given key characters are not repetitively inserted in the earlier part of the K-Box. In the present paper, the identity S-Box of RC4 is replaced by a nonidentity S-Box obtained from multiplicative polynomial inverses over Galois Fields $GF(7^3)$. The few initial K-Box elements are also obtained from some elements of multiplicative inverses and the rest are the repetition of given keys.

In Sect. 2, an overview of algebraic method to calculate multiplicative inverse over $GF(7^3)$ is given. The RC4 algorithm in brief is presented in Sect. 3. Purpose of the NIST Statistical Randomness Test Suite is briefly described in Sect. 4. The proposed new ciphering technique is presented in Sect. 5. The NIST tests are undertaken and the results are described in Sect. 6. The conclusion is in Sect. 7.

2 Algebraic Method to Find Multiplicative Polynomial Inverse Over GF(7^3)

An algebraic method to calculate the multiplicative polynomial inverses over Galois Field GF(7^3) under an irreducible polynomial [9–12] is available in literature [1]. Considering binary field ($q = 2^m$), one can use Extended Euclidean Algorithm (EEA) to calculate the multiplicative inverses of all its elemental polynomials [13–18]. The EEA is unable to find all the multiplicative inverses under an irreducible polynomial over GF(7^3), while the algebraic method, presented in [1] can do —in fact it is used here to calculate the same. In Sect. 2.1, the method is briefly discussed and its computational algorithm is presented in Sect. 2.2. An application of the algorithm to calculate multiplicative inverse of an element is shown in Sect. 2.3.

2.1 Multiplicative Polynomial Inverse Over GF(7^3)

Let $I(x) = (x^3 + a_2x^2 + a_1x + a_0)$ be a monic irreducible polynomial and one has to find multiplicative polynomial inverse of $b(x) = (b_2x^2 + b_1x + b_0)$ under this irreducible polynomial.

If $c(x) = (c_2x^2 + c_1x + c_0)$ be the multiplicative polynomial inverse then we can write,

$$[b(x)\,c(x)] \bmod I(x) = 1$$

or,

$$[(b_2x^2 + b_1x + b_0)\,(c_2x^2 + c_1x + c_0)] \bmod (x^3 + a_2x^2 + a_1x + a_0) = 1. \quad (1)$$

Solving Eq. (1), one can get the values c_2, c_1 and c_0. After few steps, the Eq. (1) can be written as follows:

$$\begin{aligned}
[\{(a_2^2b_2 &- a_1b_2 - a_2b_1 + b_0)c_2 + (b_1 - a_2b_2)c_1 + b_2c_0\}x^2 \\
&+ \{(a_1a_2b_2 - a_0b_2 - a_1b_1)c_2 + (b_0 - a_1b_2)c_1 + b_1c_0\}x \\
&+ \{(a_0a_2b_2 - a_0b_1)\,c_2 - a_0b_2c_1 + b_0c_0\}] \bmod (x^3 + a_2x^2 + a_1x + a_0) = 1.
\end{aligned}$$
$$(2)$$

Now in Eq. (2), the dividend is smaller than the divisor. Hence to satisfy the required condition of the remainder = 1, the following properties must hold.

(i)   The coefficients of $x^2 \equiv 0 \bmod 7$.
(ii) The coefficients of $x \equiv 0 \bmod 7$.
(iii) The constant part $\equiv 1 \bmod 7$.

Therefore,

$$\{(a_2^2 b_2 - a_1 b_2 - a_2 b_1 + b_0)c_2 + (b_1 - a_2 b_2)c_1 + b_2 c_0\} \bmod 7 = 0. \quad (3a)$$

$$\{(a_1 a_2 b_2 - a_0 b_2 - a_1 b_1)c_2 + (b_0 - a_1 b_2)c_1 + b_1 c_0\} \bmod 7 = 0. \quad (3b)$$

$$\{(a_0 a_2 b_2 - a_0 b_1)c_2 - a_0 b_2 c_1 + b_0 c_0\} \bmod 7 = 1. \quad (3c)$$

Note Here $GF(7^3)$ is used, and in modular arithmetic with modulus 7, the -1 is equivalent to $(-1 + 7) = 6$. Hence the $-X$ in Eqs. (3a), (3b) and (3c) can be written as $+6X$. Accordingly, the Eqs. (3a), (3b) and (3c) become

$$(d_{00} c_0 + d_{01} c_1 + d_{02} c_2) \bmod 7 = 0 \quad (4a)$$

$$(d_{10} c_0 + d_{11} c_1 + d_{12} c_2) \bmod 7 = 0 \quad (4b)$$

$$(d_{20} c_0 + d_{21} c_1 + d_{22} c_2) \bmod 7 = 1 \quad (4c)$$

where

$$d_{00} = (b_2)\%7, \ d_{01} = (b_1 + 6a_2 b_2)\%7, \ d_{02}$$
$$= (b_0 + 6a_2 b_1 + 6a_1 b_2 + a_2 a_2 b_2)\%7 \quad (5a)$$

$$d_{10} = (b_1)\%7, \ d_{11} = (b_0 + 6a_1 b_2)\%7, \ d_{12}$$
$$= (0 + 6a_1 b_1 + 6a_0 b_2 + a_1 a_2 b_2)\%7 \quad (5b)$$

$$d_{20} = (b_0)\%7, \ d_{21} = (0 + 6a_0 b_2)\%7, \ d_{22} = (0 + 6a_0 b_1 + 0 + a_0 a_2 b_2)\%7 \quad (5c)$$

All the nine d-values in Eqs. (5a), (5b) and (5c) are known. The Eqs. (4a), (4b) and (4c), i.e., $(d \times c) \bmod 7 = e$ can be solved using matrix method as,

$$c = (d^{-1} \times e)\%7. \quad (6)$$

where

$$e = \begin{pmatrix} 0 \\ 0 \\ 1 \end{pmatrix}, \quad d = \begin{pmatrix} d_{00} & d_{01} & d_{02} \\ d_{10} & d_{11} & d_{12} \\ d_{20} & d_{21} & d_{22} \end{pmatrix} \quad (7)$$

$$d^{-1} = \begin{pmatrix} id_{00} & id_{01} & id_{02} \\ id_{10} & id_{11} & id_{12} \\ id_{20} & id_{21} & id_{22} \end{pmatrix}, c = \begin{pmatrix} c_0 \\ c_1 \\ c_2 \end{pmatrix} = \begin{pmatrix} id_{02} \\ id_{12} \\ id_{22} \end{pmatrix} \quad (8)$$

While calculating d^{-1} from d-matrix, one has to ensure that the determinant det (d) is nonzero. In the event $\det(d) = 0$, the $I(x)$ is not an irreducible polynomial, rather a reducible one and d^{-1} matrix for such a case does not exist. If $\det(d)$ is nonzero for all elements, the $I(x)$ is irreducible and the multiplicative inverses of elements exist. By calculating d^{-1} from d-matrix given in Eq. (7), one can get solution for c_0, c_1, and c_2 using Eq. (8).

Now, $(b_2 x^2 + b_1 x + b_0)^{-1} = (c_2 x^2 + c_1 x + c_0) \bmod (x^3 + a_2 x^2 + a_1 x + a_0)$.

2.2 Algorithm to Find the Multiplicative Polynomial Inverses Over $GF(7^3)$

An indigenous C-program, consisting of a loop-index (ep) varying from 1 to 342 for a given irreducible polynomial (ip) and two subfunctions, is developed. The ip is converted to a polynomial using a function coeff_pol(), stored in array a[ ] and then the cal_inverse() function is called after entering the ep-loop. The cal_inverse() first calls the coeff_pol() to calculate elemental polynomial based on ep and stores them in array b[ ]. Then using the values in arrays a[ ] and b[ ], the d-matrix, given in Eq. (7), is formed and the determinant $\det(d)$ is calculated. If $\det(d) = 0$, it concludes that the current ip is not an irreducible polynomial. If $\det(d) \neq 0$, the d^{-1} is calculated whose third column is the array c[ ] shown in Eq. (8). Program algorithm is described below in pseudocode:

```
Input: Decimal equivalent of an irreducible polynomial.
Convert the ip into its septenary equivalent and store
them in an array a[] defined in eq.(1) where a_0 is the
least significant septenary digit.
For ep = 1 to 342
    Convert the ep into its septenary equivalent and
    store them in an array b[]defined in eq.(1) where b_0
    is the least significant septenary digit.
    From arrays a[] and b[] form the 3×3 d-matrix
    described in eq.(7).
    Calculate determinant of d-matrix det(d).
    If det(d)=0
        No inverse exist for current ep, hence the given
        ip is not an irreducible polynomial.
        Break.
    Otherwise
        Calculate inverse of d-matrix as d⁻¹-matrix. The
        result for the c coefficients in eq.(1) is
        obtained from eq.(8) as c₀=id₀₂,c₁=id₁₂,c₂=id₂₂
    End of for.
```

The above algorithm generates 342 multiplicative polynomial inverses for an irreducible polynomial. Under a particular irreducible polynomial, if the corresponding determinant $\det(d) \neq 0$, for a particular elemental polynomial, the algorithm calculates its multiplicative inverse; and if $\det(d) = 0$, the algorithm stops looking for further multiplicative inverses and declares that the current element (ep) has no inverse and the given ip is not an irreducible one.

2.3 Calculating Multiplicative Polynomial Inverse of $(x^2 + x + 6)$ Over $GF(7^3)$ Under an Irreducible Polynomial $(x^3 + 2x^2 + 6x + 1)$

Here the multiplicative polynomial inverse of an elemental polynomial $(x^2 + x + 6)$ is calculated under an irreducible polynomial over $GF(7^3)$ using the algebraic method.

Let the irreducible polynomial $I(x) = x^3 + a_2x^2 + a_1x + a_0 = x^3 + 2x^2 + 6x + 1$

The given polynomial $b(x) = b_2x^2 + b_1x + b_0 = x^2 + x + 6$

One have to find

$$b(x)^{-1} = c(x) = c_2x^2 + c_1x + c_0. \tag{9}$$

Here $a_2 = 2$, $a_1 = 6$, $a_0 = 1$ and $b_2 = 1$, $b_1 = 1$, $b_0 = 6$.

Using a and b values in Eqs. (5a), (5b) and (5c), one can calculate the d values as

$$d_{00} = 1\%7 = 1,\ d_{01} = 13\%7 = 6,\ d_{02} = 58\%7 = 2$$

$$d_{10} = 1\%7 = 1,\ d_{11} = 42\%7 = 0,\ d_{12} = 54\%7 = 5$$

$$d_{20} = 6\%7 = 6,\ d_{21} = 6\%7 = 6,\ d_{22} = 8\%7 = 1.$$

Following Eqs. (7) and (8), the d-matrix and its inverse d^{-1} will, respectively, be

$$d^{-1} = \begin{pmatrix} 1 & 6 & 2 \\ 1 & 0 & 5 \\ 6 & 6 & 1 \end{pmatrix},\ d^{-1} = \begin{pmatrix} 6 & 3 & 1 \\ 4 & 5 & 2 \\ 3 & 1 & 4 \end{pmatrix}.$$

Following Eq. (8), the solution of c_0, c_1 and c_2 in Eq. (9) will be obtained from d^{-1} matrix.

The solution for this problem is $\begin{bmatrix} c_0 \\ c_1 \\ c_2 \end{bmatrix} = \begin{bmatrix} 1 \\ 2 \\ 4 \end{bmatrix}.$

So one can obtain the required multiplicative inverse using Eq. (9) as

$$b(x)^{-1} = c(x) = c_2x^2 + c_1x + c_0 = 4x^2 + 2x + 1$$

Hence $(x^2 + x + 6)^{-1} = 4x^2 + 2x + 1$.

It is simple to readily verify the fact that the elemental polynomial $(4x^2 + 2x + 1)$ is indeed the multiplicative polynomial inverse of the elemental polynomial $(x^2 + x + 6)$ over $GF(7^3)$ under the irreducible polynomial $(x^3 + 2x^2 + 6x + 1)$.

3   Brief Description of RC4 Algorithm

Ron Rivest translated the shuffling concept in two stages, Key Scheduling Algorithm (KSA) and Pseudo Random Generator Algorithm (PRGA). Design procedure of 256-byte state vector S which is used as the key-pool in RC4 is very simple. And let G is the given key of length *keylen*. In KSA, a K-Box of 256-byte is created from given key and the S-Box is organized as follows:

KSA
Initialization of S vector:
for i = 0 to 255
         S[i] = i;
Generation of T vector:
for i = 0 to 255
         K[i] = G[i **mod** *keylen*]
Permutation of S vector (key mixing):
j = 0;
for i = 0 to 255 **do**
         j = (j + S[i] + K[i]) **mod** 256;
         Swap (S[i], S[j]);

After the permutation of S vector, the input key is no longer in use. Only the S vector is used in PRGA to provide a sequence of key stream Z as follows:

PRGA
i = j = 0;
while (true)
         i = (i + 1) **mod** 256;
         j = (j + S[i]) **mod** 256;
         Swap (S[i], S[j]);
         t = (S[i] + S[j]) **mod** 256;
         Z = S[t];

4   Purpose of NIST Statistical Randomness Testing Algorithms

The statistical test suite developed by NIST [19–22] is an excellent and exhaustive document to look into the various aspects of random property in a long sequence of bits. The test suite was developed to choose the Advanced Encryption Standard (AES). It is an important tool to understand and measure the randomness of Pseudo Random Number Generators (PRNG) and crypto ciphers. The NIST has documented 15 statistical tests which are well reviewed recently and available in [23]. Only the purpose of all the 15 tests are briefly noted below in Sects. 4.1–4.15.

4.1   Frequency Test

This test measures the frequencies of 0 and 1 s in the entire n-bit sequence and observes if the proportions of 0 and 1 s to n are close to 0.5.

4.2   Frequency Test Within a Block

If the first 50 % bits in the given n-bit sequence is 0 and the rest 50 % bits is 1 then the sequence would pass the frequency test mentioned in Sect. 4.1 although it is a nonrandom sequence. This test checks whether the frequencies of 0 and 1 s are uniformly distributed in the sequence.

4.3   Runs Test

Runs of length R means R number of identical consecutive bits is bounded by opposite bits. This test checks whether the frequencies of runs for various lengths of 0 and 1 s are within the limits of statistical measure.

4.4   Longest Run of Ones in a Block Test

This test measures longest run of 1 s to see if the frequencies for that appearing in the sequence are within the limit satisfying a random sequence.

4.5 Binary Matrix Rank Test

This test searches any existing repetitive patterns across the entire sequence by calculating the rank of matrices. If a matrix has full rank, then one can conclude that no repetitive pattern exists in that matrix.

4.6 Discrete Fourier Transform Test

This test checks periodic features in the sequence by undertaking Discrete Fourier Transform of each bit and calculating their peak heights. In a random sequence at least 95 % of the peak heights should be less than the threshold value of peak height.

4.7 Nonoverlapping Template Matching Test

This test finds similarity of a prespecified nonperiodic template in the given sequence in nonoverlapping manner.

4.8 Overlapping Template Matching Test

This test finds the occurrences of a prespecified template in the given sequence in overlapping manner. Through this test, one tries to detects irregularities in occurrences of any periodic template in the long bit sequence. For a template of R-bit pattern, the test accepts a sequence if the occurrences of R-runs of one lies in a specific region.

4.9 Maurer's "Universal Statistical" Test

This test concludes the compressibility of a long sequence. A sequence would be considered as random if it is not significantly compressible. This test measures the distances in terms of L-bit block numbers between L-bit templates using logarithmic function.

4.10 Linear Complexity Test

Linear Feedback Shift Register (LFSR) can produce both random and nonrandom bit sequences. For random sequence, LFSR is larger and for nonrandom sequence it is shorter. The Berlekamp–Massey Algorithm can calculate the length of LFSR from a long bit sequence. This test finds the length of LFSR for a sequence and concludes if it is random or not.

4.11   Serial Test

In a long random bit sequence, every r-bit pattern has the equal possibility of occurring. If the sequence length is n-bit, then number of occurrences of each and every r-bit overlapping patterns is approximately $n/2^r$. This test calculates the frequencies of all possible overlapping patterns in the sequence to measure the randomness of the sequence.

4.12   Approximate Entropy Test

This test measures the randomness of a long bit sequence by calculating entropy using repeating bit patterns. For a particular sequence if entropy is higher, the sequence is considered to be random. The entropy is calculated by comparing the frequencies of all possible overlapping k-bit patterns with that of $(k + 1)$-bit patterns.

4.13   Cumulative Sums Test

This test observes whether large numbers of 0 and 1 s are situated at front side or at rear side or these are mixed up equally likely across the whole bit sequence.

4.14   Random Excursions Test

This test observes whether the frequency of visits for a particular cumulative sums state in a cycle lies under the expected range of random bit sequence or not. In the test, eight states ± 1, ± 2, ± 3, and ± 4 are observed. If the visit is larger than $+4$, it is added with $+4$ state and if the visit is smaller than -4 it is added with -4 state.

4.15   Random Excursions Variant Test

This test observes the frequency of visits in a random walk to a particular state in cumulative sums manner in the given long bit sequence and calculates deviations from expected random values. In the test, 18 states $\pm1, \pm2, \pm3, \pm4, \pm5, \pm6, \pm7, \pm8$, and ±9 are observed.

5 Proposed Dynamic Ciphering Algorithm DC15

The proposed Dynamic Ciphering-15 (DC15) algorithm is an application of multiplicative polynomial inverse over $GF(7^3)$ in message encryption. The ciphering algorithm uses a PRNG producing 8-bit number sequence obtained based on the idea of random shuffling adopted in RC4. In the present paper, multiplicative inverses of an irreducible polynomial $(x^3 + 2x^2 + 6x + 1)$ are obtained and their larger part is used in the S-Box and a smaller part, in the initial part of K-Box. In DC15, the identity S-Box is replaced by a nonidentity S-Box that sequentially takes decimal equivalent of 256 multiplicative inverses less than $(514)_7$. The decimal equivalent of other 87 multiplicative inverses equal to and greater than $(514)_7$ are complemented at bit level and its decimal equivalent is sequentially put in the K-Box as the first initial 87 entries. The rest 169 spaces of the K-Box are filled by the given key following the RC4 algorithm. As the initial S-Box is not an identity S-Box and the initial K-Box elements do not contain the given key elements, the present algorithm is cryptographically better than the conventional RC4 and the initial PRGA bytes created by DC15 would not carry the bias of the given key as mentioned in [4, 6]. Subsequently, a dynamic permutation is introduced among the randomly generated key-bytes before they are XORed sequentially with the message bytes. The ciphering algorithm is as follows:

```
While true, do
    key[0] = randomly generated key-byte.
    setlen = key[0] mod 15 + 1. // setlen: length of a set
                // of bytes to be permuted randomly at a time.

    If(setlen > 1)
        For loop = 1 to setlen-1
            key[loop] = randomly generated key-byte.

    For loop = 0 to setlen-1
        msg[loop] = sequentially read message-byte from
                    file to be encrypted.

    For loop = 0 to setlen-1

        cipher = msg[loop]^key[setlen - 1 - loop]
                            // cipher-byte is generated.
        Write cipher-byte in output file.
End of while.
```

Position of randomly generated
key-bytes, before permutation.

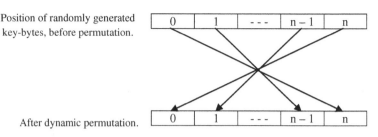

After dynamic permutation.

Fig. 1 Dynamic permutation of randomly generated key-bytes

The dynamic permutation of the randomly generated key-bytes can be explained by a graphical presentation as shown in Fig. 1. Let n be the number of generated key-bytes, then first key-byte goes to the nth position, second goes to the $(n - 1)$th position, so on and so forth.

It is intended to compare the randomness of cipher bytes produced by DC15 with that produced by RC4 from statistical randomness perspective. The statistical randomness tests are undertaken on the two algorithms—their results are shown and discussed in Sect. 6.

6 Results of Comparative Statistical Randomness Testing

The motivation of incorporating the NIST test suite is to study and compare the randomness of the output cipher obtained using the proposed DC15 with that obtained using the RC4—both on the same message block. A calculation technique is presented in NIST document in which the χ^2-value coupled with the degrees of freedom is transformed to a P-value. Thereby, it sets a passing criterion of P-value ≥ 0.01 (significance level). Using all the P-values obtained for a particular test, NIST also mentioned necessary statistical procedures to compute the proportion of passing and the uniformity of P-values.

It may be noted that for each of the two algorithms, 300 different keys each of 16-bytes are used to encrypt a plain text of 1,68,960 bytes. Each algorithm generates 300 different cipher bit-sequences each of 13,51,680 bits obtained by XORing necessary random key-bytes with text bytes. Following the recommendation of NIST, the minimum bit lengths as required for respective tests are shown in Table 1.

In estimating the randomness of a particular NIST test, it is necessary to consider two checking parameters: (1) threshold value or the Expected Proportion Of Passing (EPOP) and (2) P-value of P-values (POP). These two are explained in Sects. 6.1 and 6.2, respectively. Following Tables 2 and 3, the result of DC15 is compared with that of RC4 and is presented in Table 4.

Table 1 Minimum required lengths and used lengths of bit sequence for 15 statistical tests

Test no.	Test name	Length of bit sequence (n)	
		Minimum requirement	Used in present software
1	Frequency test	100	1,342,400
2	Frequency test within a block	9,000	1,342,400
3	Runs test	100	1,342,400
4	Longest run of ones in a block	128	1,342,400
5	Binary matrix rank test	38,912	1,342,400
6	Discrete fourier transform test	1,000	13,424
7	Nonoverlapping template test	10,48,576	1,342,400
8	Overlapping template test	10,00,000	1,342,400
9	Maurer's "universal statistical" Test	13,42,400	1,342,400
10	Linear complexity test	10,00,000	1,342,400
11	Serial test	10,00,000	1,342,400
12	Approximate entropy test	100	1,342,400
13	Cumulative sums (cusum) test	100	13,424
14	Random excursions test	10,00,000	1,342,400
15	Random excursions variant test	10,00,000	1,342,400

6.1 EPOP and Observed Proportion of Passing (OPOP)

To estimate EPOP of a particular test, a large sample size of sequences of bits obtained randomly as the output by an algorithm is to be considered. Now let the sample size of sequences of bits be m and a test generates a single P-value from each and every sequence, then EPOP would be calculated as

$$\text{EPOP} = (1 - \alpha) - 3\sqrt{\frac{\alpha(1 - \alpha)}{m}} \tag{10}$$

Here in Eq. (10), α is the significance level. The sample size m should be larger than the inverse of the significance level α. For $m = 300$ and $\alpha = 0.01$, EPOP = 0.972766. A particular test producing one P-value can be considered to be statistically successful, if at least its 292 P-values out of its 300 P-values do pass the test. For any test producing n number of P-values, $(m \times n)$ should be considered instead of m in Eq. (10). With the same values of α and m, the EPOP for Random Excursions test with $n = 8$ is 0.983907. If at least 2362 P-values out of the total $300 \times 8 = 2400$ P-values pass the test, then this test is to be considered as statistically successful. The OPOP of a particular test is defined as the ratio of passing P-values with total P-values. The status for proportion of passing would be considered as a success if OPOP $\geq$ EPOP for a particular test.

6.2   P-value of P-values (POP) and Uniformity of Distribution of P-values

For a particular test, the uniformity of distribution of P-values can be realized from its obtained P-values. The P-value data for a particular test are divided in 11 groups between 0 and 1 and such data for all tests are shown in Tables 2 and 3 for RC4 and DC15, respectively.

Depending on the result of a P-value for a particular test, the count of an appropriate group is increased in which the particular P-value belongs. If P-value <0.01, then it will be considered as unsuccessful. The entry in column 1 of Tables 2 and 3 indicates the numbers of unsuccessful P-values. For estimating χ^2-deviation of distribution of P-values, the first two P-values are merged in one group and the rest in nine groups, thereby 10 groups of P-values are considered. The χ^2-deviation of distribution of P-values is computed as

$$\chi^2 = \sum_{i=1}^{10} \frac{\left(s_i - \frac{m}{10}\right)^2}{m/10} \tag{11}$$

where S_i is the number of P-values in ith group and m is sample size. If a particular test produces n number of P-values, then $m \times n$ should be considered instead of m in Eq. (11). Here the degrees of freedom $v = 9$ and the sample size is the number of cipher texts considered for NIST testing. The two parameters in $\Gamma(a, x)$ function are calculated as $a = v/2$ and $x = \chi^2/2$ and the corresponding POP is obtained as

Table 2 Frequency distribution of P-values of RC4

Test no.	0.0–0.01	0.01–0.1	0.1–0.2	0.2–0.3	0.3–0.4	0.4–0.5	0.5–0.6	0.6–0.7	0.7–0.8	0.8–0.9	0.9–1.0
1	15	54	38	38	28	30	22	14	19	20	22
2	99	112	40	21	11	7	1	4	4	1	0
3	10	21	22	26	39	35	32	24	29	28	34
4	4	20	27	30	18	42	32	37	24	38	28
5	41	80	46	26	28	22	15	8	13	13	8
6	2	31	31	36	23	35	20	34	34	34	20
7	2	28	35	32	28	24	22	35	29	35	30
8	2	27	17	31	31	43	36	31	27	28	27
9	300	0	0	0	0	0	0	0	0	0	0
10	1	19	35	41	28	39	27	30	21	29	30
11	300	0	0	0	0	0	0	0	0	0	300
12	300	0	0	0	0	0	0	0	0	0	0
13	5	59	49	74	77	64	54	52	63	53	50
14	35	208	225	246	250	239	220	218	232	261	266
15	37	474	522	487	534	529	527	575	564	588	563

$$POP = 1 - \frac{\Gamma(a,x)}{\Gamma(a,\infty)} \tag{12}$$

The *P*-values are considered as uniformly distributed if Eq. (12) gives POP ≥ 0.0001.

6.3   Results of Comparative Study Between RC4 and DC15

The test-wise results of statistical tests on RC4 are given in Table 2 and that for DC15, in Table 3. The OPOP for a particular test is the ratio of sum of the last ten columns to the total sum of 11 columns. If OPOP ≥ EPOP, the particular test indicates that the data set is statistically random. The uniformity of distribution of *P*-values obtained for a particular test is indicated by *P*-value of *P*-values (POP) computed using Eq. (12)—these can also be derived from Tables 2 and 3. The distribution is considered to be uniform if POP ≥ 0.0001. Following the procedures stated above, the test-wise EPOP, OPOP, and POP data for the algorithms RC4 and DC15 are calculated for all the 15 tests and are shown in Table 4.

From OPOP data shown in Table 4, two algorithms are observed to be at the same footing—RC4 has better OPOP for five tests (7, 10, 13, 14, 15), while DC15 has better OPOP also for five tests (1, 2, 3, 4, 5). The rest five tests give same OPOP for both the algorithms. From POP data shown in Table 4, it is observed that nine tests (1, 4, 5, 7, 8, 10, 13, 14, 15) of DC15 indicate uniform distribution of

Table 3 Frequency distribution of *P*-values of DC15

Test no.	0.0–0.01	0.01–0.1	0.1–0.2	0.2–0.3	0.3–0.4	0.4–0.5	0.5–0.6	0.6–0.7	0.7–0.8	0.8–0.9	0.9–1.0
1	9	58	39	32	24	31	14	28	20	25	20
2	97	117	43	19	14	5	3	1	0	1	0
3	4	33	28	38	29	26	28	32	37	28	17
4	2	25	31	31	35	27	36	25	26	29	33
5	38	77	56	31	33	17	15	11	4	7	11
6	2	32	40	25	18	31	23	31	34	30	34
7	5	32	27	27	33	22	31	34	29	27	33
8	2	27	37	20	26	34	25	32	31	32	34
9	300	0	0	0	0	0	0	0	0	0	0
10	5	21	34	31	33	33	32	29	30	28	24
11	300	0	0	0	0	0	0	0	0	0	300
12	300	0	0	0	0	0	0	0	0	0	0
13	9	52	59	76	60	63	59	70	42	52	58
14	41	211	246	232	238	233	242	259	245	240	213
15	70	501	530	520	532	531	544	562	536	542	532

Table 4 Observed proportion of passing (OPOP) and P-value of P-values (POP)

Test no.	EPOP	Observed proportion of passing (OPOP)		P-value of P-values (POP)	
		RC4	DC15	RC4	DC15
1	0.972766	0.950000	0.970000	1.399834e−12	9.783167e−11
2	0.972766	0.670000	0.676667	2.364251e−265	5.984242e−276
3	0.972766	0.966667	0.986667	5.075122e−01	2.058966e−01
4	0.972766	0.986667	0.993333	6.023866e−02	8.831714e−01
5	0.972766	0.863333	0.873333	3.616519e−69	4.931339e−68
6	0.972766	0.993333	0.993333	1.986900e−01	1.986899e−01
7	0.972766	0.993333	0.983333	7.265031e−01	7.531852e−01
8	0.972766	0.993333	0.993333	1.480942e−01	5.612272e−01
9	0.972766	0.000000	0.000000	0.000000e+00	0.000000e+00
10	0.972766	0.996667	0.983333	1.199735e−01	9.558347e−01
11	0.977814	0.500000	0.500000	0.000000e+00	0.000000e+00
12	0.972766	0.000000	0.000000	0.000000e+00	0.000000e+00
13	0.977814	0.991667	0.985000	9.278355e−02	1.782783e−01
14	0.983907	0.985417	0.982917	3.384789e−01	7.498835e−01
15	0.985938	0.993148	0.987037	5.634658e−02	9.077180e−01

P-values, while only three tests (2, 3, 6) of RC4 indicate uniform distribution of P-values and the rest three tests of both the algorithms show the same result.

Regarding the uniform distribution of P-values, one can correlate the POP value shown in Table 4 with a corresponding visual histogram obtained from the right data of Tables 2 and 3. From Table 4, the test no. 10 of DC15 is seen as the best POP—the same data is displayed in a corresponding histogram in Fig. 2. The uniformity of the P-value distribution is visually evident. The best POP of RC4 is for test 7 for which a histogram is shown in Fig. 3. There are ten columns in the histograms where each column indicates the count of P-values lying within a predefined range as 0.0–0.1 for first column, 0.1–0.2 for second column, and so on.

Fig. 2 Histogram for test no. 10 of DC15 (POP: 9.558347e−01)

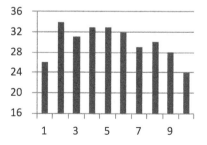

Fig. 3 Histogram for test no. 7 of RC4 (POP: 7.265031e−01)

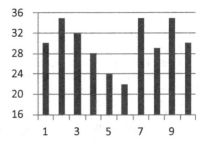

7 Conclusion

This paper shows a practical implementation of multiplicative polynomial inverses over GF(7^3) in cryptographic application. In the proposed DC15 algorithm, the initial S-Box is filled by multiplicative inverses of all its elemental polynomials under one of the 112 irreducible polynomials over GF(7^3). Initial part of the K-Box is also filled by multiplicative inverses of some other elemental polynomials. The comparative study of DC15 and RC4 based on respective NIST data indicates that DC15 is better than RC4 from randomness point of view. By incorporating multiplicative polynomial inverses in K-Box, DC15 removes the bias of the given key from its initial random key-bytes unlike RC4.

Acknowledgments We express our gratitude toward the DST, New Delhi and the TEQIP (Phase-II), University of Calcutta for providing financial support, respectively, to the first author and the second author. We are also indeed thankful to the Head of the Department of Radio Physics and Electronics, University of Calcutta for providing necessary infrastructural facilities to undertake the research activities.

References

1. Zaman, J.K.M.S., Ghosh, R.: Multiplicative polynomial inverse over GF(7^3): crisis of EEA and its solution. Appl. Comput. Sec. Syst. **2**, 87–107 (2014). (Springer)
2. Stallings, W.: Finite fields. In: Cryptography and Network Security Principles and Practices, 4th ed., pp. 95–133. Pearson Education, Delhi (2008)
3. Forouzan, B.A., Mukhopadhyay, D.: Mathematics of cryptography. In: Cryptography and Network Security, 2nd edn., pp. 15–43. TMH, New Delhi (2011)
4. Roos, A.: A Class of Weak Keys in the RC4 Stream Cipher (Sept. 1995)
5. Fluhrer, S., Mantin, I., Shamir, A.: Weakness in the key scheduling algorithm of RC4. In: Proceedings of International Workshop on Selected Areas in Cryptography, Berlin Heidelberg, LNCS 2259, pp. 1–24 (2001)
6. Paul, B., Preneel: A new weakness in the RC4 Keystream. Generator and an approach to improve the security of the cipher. In: Proceedings of Fast Software Encryption, Berlin, LNCS 3017, pp. 245–259 (2004)
7. Maitra, S., Paul, G.: Analysis of RC4 and proposal of additional layers for better security margin. In: Proceedings of Indocrypt, IIT Kharagpur, LNCS 5365, pp. 27–39 (2008)

8. Paul, G., Maitra, S.: RC4 Stream Cipher and Its Variants. Chapman & Hall/CRC, Boca Raton (2012)
9. Stinson, D.R.: The RSA cryptosystem and factoring integers. In: Cryptography Theory and Practice, 3rd edn., pp. 161–232 Chapman & Hall/CRC, Boca Raton (2006)
10. Knuth, D.E.: The Art of Computer Programming Seminumerical Algorithms, 3rd edn., Vol. 2. Pearson Education, Upper Saddle River (2011)
11. Church, R.: Tables of irreducible polynomials for the first four prime moduli. Ann. Math. **36** (1), 198–209 (1935)
12. Lidl, R., Niederreiter, H.: Finite Fields, Encyclopedia of Mathematics and Its Applications, Vol. 20. Addison-Wesley Publishing Company (1983)
13. Arguello, F.: Lehmer-based algorithm for computing inverses in Galois fields GF(2^m). Electron. Lett. IET J. Mag. **42**(5), 270–271 (2006)
14. Yan, Z., Starwate, D.V.: New systolic architectures for inversion and division in GF(2^m). IEEE Trans. Comput. **52**(11), 1514–1519 (2003)
15. Hasan, M.A.: Double-basis multiplicative inversion over GF(2^m). IEEE Trans. Comput. **47**(9), 960–970 (1998)
16. Guo, J.H., Wang, C.L.: Systolic array implementation of euclid's algorithm for inversion and division in GF(2^m). IEEE Trans. Comput. **47**(10), 1161–1167 (1998)
17. Brunner, H., Curiger, A., Hofstetter, M.: On computing multiplicative inverses in GF(2^m). IEEE Trans. Comput. **42**(8), 1010–1015 (1993)
18. Wang, C.C., Truong, T.K., Shao, H.M., Deutsch, L.J., Omura, J.K., Reed, I.S.: VLSI architecture for computing multiplications and inverses in GF(2^m). IEEE Trans. Comput. **C-34** (8), 709–717 (1985)
19. Rukhin, A., Soto, J., et al.: A statistical test suite for random and pseudorandom number generators for cryptographic applications. NIST, US, Technology Administration, U.S. Department of Commerce (2010)
20. http://csrc.nist.gov/publications/nistpubs/800-22-rev1a/SP800-22rev1a.pdf
21. Rukhin, A., Soto, J., et al.: A statistical test suite for random and pseudorandom number generators for cryptographic applications. NIST, Technology Administration, U.S. Department of Commerce (2008)
22. http://csrc.nist.gov/groups/ST/toolkit/rng/documentation_software.html
23. Zaman, J.K.M.S., Ghosh, R.: Review on fifteen statistical tests proposed by NIST. Int. J. Theor. Phys. Crypt. **1**, 18–31 (2012)

On Preventing SQL Injection Attacks

Bharat Kumar Ahuja, Angshuman Jana, Ankit Swarnkar and Raju Halder

Abstract In this paper, we propose three new approaches to detect and prevent SQL Injection Attacks (SQLIA), as an alternative to the existing solutions namely: (i) Query Rewriting-based approach, (ii) Encoding-based approach, and (iii) Assertion-based approach. We discuss in detail the benefits and shortcomings of the proposals w.r.t. the literature.

Keywords SQL injection attacks · Query rewriting · Encoding · Assertion checking

1 Introduction

SQL injection is an attack in which malicious SQL code is inserted or appended into database application through user input parameters that are later passed to a back-end SQL server for parsing and execution [11]. The growing popularity and intense use of database-driven web applications in today's web-enabled society make them ideal target of SQL Injection Attacks (SQLIA). The number of SQLIA

B.K. Ahuja (✉) · A. Jana · A. Swarnkar · R. Halder
Indian Institute of Technology Patna, Patna, India
e-mail: bharat.cs10@iitp.ac.in

A. Jana
e-mail: ajana.pcs13@iitp.ac.in

A. Swarnkar
e-mail: ankitswarnkar.cs10@iitp.ac.in

R. Halder
e-mail: halder@iitp.ac.in

© Springer India 2016
R. Chaki et al. (eds.), *Advanced Computing and Systems for Security*,
Advances in Intelligent Systems and Computing 395,
DOI 10.1007/978-81-322-2650-5_4

has increased rapidly in recent years, and it has become a predominant type of attacks. Consider the following PHP script from [4]:

```
\\ Connect to the database
    $conn = mysql_connect("localhost", "username", "password");

\\ Dynamically building sql statement with user input
    $query="SELECT * FROM Products WHERE Price<'$_GET["val"]'".
                                "ORDER BY ProductDescription";
\\ Executing the query against the database
    $result = mysql_query($query);
```

The URL "http://www.victim.com/products.php?val=100" is used to view all products with cost less than $100. However, on providing malicious input '**100**' **OR '1' = '1'** and the corresponding URL http://www.victim.com/products.php?val = '**100**' **OR '1' = '1'**, the dynamically created SQL statement "SELECT * FROM Products WHERE Price < '100' OR '1' = '1' ORDER BY ProductDescription" extracts all product information as the WHERE clause evaluates to true always.

In addition to the above tautology-based SQL injection attacks, there exist various other forms of attacks with various attacker intents. For instance, Union Query, Piggy-Backed Query, Stored Procedures, etc. [11].

The primary reason behind SQL injection is the direct insertion of code into parameters that are finally concatenated with SQL commands and executed. When Web applications fail to properly sanitize the inputs, it is possible for an attacker to alter the construction of back-end SQL statements, which may lead to a catastrophe. Web applications along with databases, therefore, require not only careful configuration and programming to ensure data security, but also need effective and efficient protection techniques to prevent SQL injection attacks.

A wide range of prevention and detection techniques are proposed to address the SQL injection problems [4, 11]. According to the best of our knowledge, no existing approach can guarantee a complete safety. For example, many proposed approaches, like AMNESIA—a model-based technique [9], taint-based technique [17], intrusion detection System [22], etc., have large number of false positive alarm. The static code checker [7] is used to prevent tautology-based attack only. The approach of instruction set randomization [2] is unable to prevent many types of SQLIA like illegal/logically incorrect query, stored procedures, alternate encodings, etc. It is extremely difficult to apply defensive coding practices [8] for all the sources of inputs because it results high rate of false positive (e.g., in case of input O'Brian).

In this paper, we propose three new directions to detect and prevent SQLIA, as alternative solutions. These are (i) query rewriting-based approach, (ii) encoding-based approach, and (iii) assertion-based approach.

The main objective of first approach is to mitigate the effect of concatenation operation during query generation. The objective of the second approach is to mitigate the mal-effect of bad input on query semantics. The assertion-based technique is a state verification technique which can be performed either at application level or at database level. We provide a comparative study among our proposed approaches based on security guaranty, usability, and applicative point of view.

The structure of the paper is as follows: Sect. 2 discusses related works in the literature. Section 3 describes and formalizes the problem. We introduce our proposed approaches in Sect. 4. Section 5 represents the comparative study and complexity analysis. Section 6 concludes the paper.

2 Related Works

In [10], authors proposed a model-based technique to prevent SQLIA that combines both static and dynamic analyses. In static phase, the approach builds character-level Nondeterministic finite automaton (NFA) model for each hotspot. All dynamically generated queries, during runtime, are checked against the corresponding model and violation of the models is reported as SQLIA. However, the success of this approach is dependent on the accuracy of its static model, and may results both false positives and false negatives. Other approaches on static models considering grammars and parse-trees are proposed in [3, 21].

Defensive coding practice [8] is a way to prevent the SQL injection vulnerabilities. But it is extremely difficult to apply for all sources of input, because in many applications SQL-keywords, operators can be used to express normal text entry, formulas or even names (e.g., O'Brian).

In [2], authors proposed Instruction set rndomization approach and introduce the SQLrand tool. It provides a framework that allows developers to create queries using randomized instructions instead of normal SQL keywords. The attacker is not aware about that randomize instruction. Thus if any malicious user attempts SQL injection attack would immediately be thwarted. However, it cannot cover all types of SQLIA. It is unable to detect or prevent the many types of SQLIA like illegal/logically incorrect query, stored procedures, alternate encodings, etc., and also it imposes a significant infrastructure overhead.

In [12], authors introduced an obfuscation-based approach to detect the presence of possible SQLIA. Obfuscation removes the need of concatenations operation and treats the atomic formulas in the condition part of the queries as independent from each other; whereas the dynamic phase, after merging the user inputs in the obfuscated atomic formulas, verifies the presence of possible SQLIA.

CANDID [1] is a code transformation-based approach which aims to construct programmer-intended query structure. In order to construct the intended query, the approach runs an application on a set of candidate inputs that are self-evidently nonattacking. However, false positive is the prime limitation of this approach.

In [14], authors proposed an idea to identify various types of SQLIA and to mitigate such attacks by redefining code-injection attacks on outputs (CIAOs). Precisely, detecting CIAOs basically follows the dynamic white box mechanisms which are based on dynamic taint analysis. The primary limitation is that determining whether an application is vulnerable to CIAOs requires the knowledge about which input symbols are propagating to the output program.

DIGLOSSIA [20] is a run time tool which performs a dual parsing to compare the shadow query (which must not be tainted by user inputs) with the actual application generated query, and ensures that the query issued by the application program does not contain injected code. However, this does not consider all sources of inputs (e.g. it does not permit any input to be used as a part of the code in the query).

Another two techniques, SQL DOM [16] and safe query object [6] are used for encapsulation of database queries to provide a safe and reliable way to access databases. The main limitation of this technique is that they require developers to learn and use a new programming paradigm or query-development process.

Several automated or semiautomated tools for detection and prevention of SQLIA are already developed. For instance, AMNESIA [9], SQLCheck [21], SQLGuaed [11], SQLrand [2], WebSSARI [18], JDBC-Checker [7], etc.

3 Problem Description

The characteristics of dynamic web applications are

- They receive inputs as strings during run time.
- They build SQL query strings by performing concatenation operation between SQL constructs and input strings.
- Bad input, after concatenation, may be treated as a part of SQL control constructs, leading to SQL Injection Attacks.

Formally, this can be defined as follows [21]: Let Σ be an alphabet. A web application P: $(\Sigma^* \times \Sigma^* \times \ldots \times \Sigma^*) \rightarrow \wp(\Sigma^*)$ is defined as a mapping from a set of user inputs over an alphabet Σ to a set of query strings of Σ. Given a set of SQL substrings $\{s_1, s_2, \ldots s_n\}$ and a set of input strings $\{i_1, i_2 \ldots i_m\}$, P generates a query string (using concatenation operation) $Q = q_1 + q_2 + \ldots + q_l$, where for $1 \leq j \leq l$:

$$q_j = \begin{pmatrix} i & \text{where } i \in \{i_1, i_2, \ldots, i_m\} \\ s & \text{where } s \in \{s_1, s_2, \ldots, s_n\} \end{pmatrix}$$

Given a query string Q, it can be divided into two parts: data-part and control-part. Let us denote $Q = \langle \{d_1, d_2, \ldots, d_p\}, \{c_1, c_2, \ldots, c_q\} \rangle$ where

$\{d_1, d_2, \ldots, d_p\}$ is the set of data-parts and $\{c_1, c_2, \ldots, c_q\}$ is the set of control-parts. SQL Injection attack occurs if $\{c_1, c_2, \ldots, c_p\} \cap \{i_1, i_2, \ldots, i_m\} \neq \phi$; that is, when any input string is treated as SQL control construct in Q.

4 Proposed Approaches

In this section, we propose three possible solutions as an alternative to the existing ones.

4.1 Query Rewriting Approach

In database-enabled web applications, SQL Queries are, in general, constructed by concatenating input strings directly from the users. This helps attackers to somehow manage and modify the query structures by providing malicious SQL keywords in the input strings. In order to mitigate the use of concatenation operation, we propose a query-rewriting approach which transforms the insecure web application into semantically secure version. The proposed approach has following two main phases:

- Static Phase
  - Insert user inputs into a database table, called "Input" table.
  - Replace the concatenation operation of user inputs with other SQL constructs by an equivalent SELECT query which selects inputs from the table "Input."
- Dynamic phase
  - Parse the INSERT statement after merging inputs to check the correctness of syntax.

Static Phase
In this phase, all the inputs of the web application are inserted into a database table, called "Input" table, using INSERT statement. The objective is to mitigate the bad effect of input data by treating them as a part of the database and to remove the necessity of concatenation operation using them during the query formation later on.

The generic structure of the "Input" table is as follows: Let the web application P involves n queries $\{Q_i \mid i = 1, \ldots n\}$. Suppose each query Q_i accepts m user inputs $S_I^{Q_i} = \{I_{ij} \mid j = 1, \ldots m\}$. The structure of "Input" table is

INPUT (QID *integer*, IID *integer*, u_input *varchar*)

where QID, IID represent program point i and input id j, which together uniquely identify jth input I_{ij} in ith query Q_i—hence forms primary key.

Static Analyzer Let l_i: Q_i ($\{I_{ij}|j = 1, \ldots, m\}$) denotes that query Q_i is formed using concatenation operation with user inputs $\{I_{ij}|j = 1, \ldots, m\}$ at program point l_i. The analyzer scans the web application and performs the following operations:

1. Identify $l_i : Q_i\left(\langle I_{ij}|j = 1, \ldots, m\rangle\right)$ for $i = 1, \ldots n$.

2. Before each l_i, add an INSERT statement which inserts all inputs $\{I_{ij}|j = 1, \ldots m\}$. For example, consider the following query:

"SELECT eid FROM emp WHERE login=' " +slogin+ "᾿ AND pass=' " +spass+ "᾿"

The analyzer adds two INSERT statements before the query, assuming it is at program point 1 as follows:

"INSERT INTO INPUT(GID, IID, u_input) VALUES ('1', '1', ' " +slogin+ " ')"

"INSERT INTO INPUT(GID, IID, u_input) VALUES ('1', '2', ' " +spass+ " ')"

"SELECT eid FROM emp WHERE login=' " +slogin+ "᾿ AND pass='" +spass+ "᾿"

3. Convert the query Q_i by semantically equivalent version where each concatenation operation and input variables are replaced by corresponding SELECT query accessing the same values from the "Input" table.

For example, the query mentioned just before is transformed as

"INSERT INTO INPUT(GID, IID, u_input) VALUES ('1', '1', ' " +slogin+ " ')"

"INSERT INTO INPUT(GID, IID, u_input) VALUES ('1', '2', ' " +spass+ " ')"

SELECT eid FROM emp WHERE login = (SELECT u_input FROM INPUT WHERE QID = 1 AND IID = 1) AND pass = (SELECT u_input FROM INPUT WHERE QID = 1 AND IID = 2)

Observe that a basic filtering is applied on inputs to filter the presence of any meta-character which is treated as control-character of INSERT statement. For instance, "is replace by."

Dynamic Phase

In the modified version of web application, it is observed that only INSERT statement is vulnerable to SQLIA. Attacker may try to change the behavior of INSERT statement through malicious inputs. A dynamic analyzer will check the correctness of the syntax of INSERT statement after merging inputs and before issuing to the database. For this purpose, we define the following grammar for INSERT statement:

```
Ins_ stmt ::= INSERT INTO E S
        E ::= Identifier
        S ::= ( attr X val )
        X ::= , attr X val , | ) VALUES (
     attr ::= Identifier
      val ::= ' id '
       id ::= (string | number | com | meta)* | null
Identifier ::= letter(letter | digit)*
      com ::= = | > | < | ⩾ | ⩽ | ! =
     meta ::= ' | ' | – | ;
```

Observe that these production rules check the number of inputs with the number of attributes in the INSERT statement. We can also add a mechanism for type checking, in addition.

Illustration with Example

Let us consider the database and web application depicted in Fig. 1a, b respectively.

Static Phase As discussed, the static analyzer adds INSERT statement aiming at inserting all inputs into the "Input" table and replaces the input strings along with the concatenation operation by equivalent SELECT queries accessing data from "Input" table. The result is shown in Fig. 1c.

Dynamic Phase Consider the code in Fig. 1c and the inputs: user and x'OR 1 = 1: After merging the inputs, the INSERT statement is

3a. INSERT INTO INPUT (QID, IID, u_input) VALUES ('3', '1', ' user ')

3b. INSERT INTO INPUT (QID, IID, u_input) VALUES ('3', '2', ' x''OR 1=1 ')

The dynamic analyzer will check the syntax w.r.t. the grammar depicted before. In this case the INSERT statements parse successfully.

Observe that the execution of inner queries select the inputs from "Input" table and yield the following:

SELECT eid FROM loginfo WHERE login = user AND pass = x'OR 1 = 1

As there is no credential (user and x'OR1 = 1) in the table "loginfo," the execution finally results into "NO ROW SELECTED."

4.2 Encoding-Based Approach

As we know any modification of query by attacker inputs using concatenation operation is only possible if both the query languages and the input strings use same

eid	login	pass
1	Admin	Admin
2	user	1234
3	O'Brian	user1234

eid	ename	B_ sal	GP
1	Raja	1600	350
2	Tapan	800	280
3	Tamal	2600	500

(a) Database (Tables "loginfo" and "salinfo")

---

1. start;

2. Stmt = DriverManager.getConnection(" jdbc:mysql://172.16.1.16:1115/
demo", "scott", "tiger").createStatement();

3. Resultset rs1 = Stmt.executeQuery("SELECT eid FROM loginfo WHERE login = ' "+ slogin +" ' AND
pass = ' "+ spass +" ' ");

4. Resultset rs2 = Stmt.executeQuery("SELECT ename FROM salinfo WHERE B_sal> GP + ' "+vsal+" ' ");

5. Resultset rs3 = Stmt.executeQuery("SELECT ename FROM salinfo WHERE GP > ' "+ com +" ' *
' "+ rate +" ' ");

6. stop;

(b) Web application

---

1. start;

2. Stmt = DriverManager.getConnection(" jdbc:mysql://172.16.1.16:1115/
demo", "scott", "tiger").createStatement();

3a. String str = "INSERT INTO INPUT (GID, IID, u_input) VALUES ('3', '1', ' "+slogin+" ')"
int i = check(str);

3b. String str1 = "INSERT INTO INPUT (GID, IID, u_input) VALUES ('3', '2', ' "+spass+" ')"
i = check(str1);

3. Resultset rs1 = Stmt.executeQuery("SELECT eid FROM loginfo WHERE login= (SELECT u_input FROM INPUT
WHERE GID = 1 AND IID = 1) AND pass = (SELECT u_input FROM INPUT WHERE GID = 1 AND IID = 2) ");

4a. String str2 = "INSERT INTO INPUT (GID, IID, u_input) VALUES ('4', '1', ' "+vsal+" ')"
i = check(str2);

4. Resultset rs2 = Stmt.executeQuery("SELECT ename FROM salinfo WHERE B_sal> GP + (SELECT u_input FROM
INPUT WHERE QID = 2 AND IID = 1)");

int mul = rate * com ;
5a. String str3 = "INSERT INTO INPUT (GID, IID, u_input) VALUES ('5', '1', ' "+mul+" ')"
i = check(str3);

5. Resultset rs3 = Stmt.executeQuery("SELECT ename FROM salinfo WHERE GP > (SELECT u_input FROM
INPUT WHERE QID = 3 AND IID = 1)");

6. stop;

(c) Modified web application

QID	IID	u_input
3	1	user
3	2	x'OR1=1
4	1	1'OR1=1
5	1	200

(d) Dynamic phase (Table "INPUT")

Fig. 1 An example

alphabet. What if we change the alphabet of input strings different from the SQL? This mitigates the vulnerable effect of inputs when directly concatenated to the original SQL statements.

Let U be the alphabet of user inputs which may overlap with the alphabet of SQL characters. Let Σ_{new} be a new alphabet different from SQL characters. We define the following function F which encodes any symbol of U into a string of Σ_{new}^*:

$$F : U \rightarrow \Sigma_{new}^*$$

and component-wise distributive property

$$F(x_1 x_2 \ldots x_n) = F(x_1)F(x_2)\ldots F(x_n)$$

As a trivial example, let us consider the binary alphabet $\Sigma_{new} = \{0, 1\}$. Given a string "abc," the corresponding binary encoded representation of it in Σ_{new} is obtained by

$$F(abc) = F(a)\ F(b)\ F(c) = 011000010110001001100011$$

As an example, consider the following query:

"SELECT eid FROM loginfo WHERE login=' " +Slogin+" ' AND pass=' " +Spass+" ' "

Consider the inputs of Slogin and Spass are: 'OR 1=1 -- and password, respectively. The encoded version of this inputs are

$$F(' \ OR \ 1=1 \ --) = 01010101010101010101010011001010001$$
$$F(password) = 10010111010101001010001010010101 0$$

Instead of merging the original inputs to the original query, we merged the encoded representation in the language different from the SQL characters, as depicted below:

SELECT eid FROM loginfo WHERE login ='01010101010101010101010011001010001' AND

pass = '01010101010101010101010011001010001'

Observe that the semantics of Q is not changed due to merging of encoded inputs, leading to a mitigation of the injection attacks. However, the issue is that WHERE condition produces false positives as values are in a new language.

4.3 Extending Encoding-Based Approach to Database Spplications

We now extend the encoding-based approach to the case of database applications, and we discuss the following two major issues:

1. **Data storage and string comparison**: Consider the table "loginfo" in Fig. 1a. Suppose the administrator wants to login by providing login = Admin and pass = Admin. Appending inputs in the encoded form results into the following:

SELECT eid FROM loginfo WHERE login ='0110000101110101010110110001010001'

AND pass = '0110000101110101010110110001010001'

where $F(\text{Admin}) = 0110000101110101011011001010001$. When string matching operation is performed, as an enhancement, we adopt either of the followings:

(a) *Encoded databases*: To enable the matching operation, we store encoded values of strings in the databases rather than actual strings, and we decode it to its original form after performing all operations at encoded level. The encoded version of the table "loginfo" (in Fig. 1a) is

Login(varbinary)	Pass(varbinary)
0110000101110101011011001010001	0110000101110101011011001010001
1001001001010100101001010000010	10101010110010101011001010101000

Observed that encoded form of database may occupy more storage space than the original one and may take more time on performing database operations. Therefore, the technique might be suitable for small database systems.

(b) *Conversion on-the-fly:* Storing values into the database in raw binary format does not seem convincing w.r.t. time and space complexity point of view, and it can take too much overhead in decoding. As a remedy, the database data can only be encoded whenever required by the quires to compare with the encoded inputs as shown in the following example.

"SELECT eid FROM loginfo WHERE databaseEncode(login)='"+

applicationEncode(Slogin)+"'AND databaseEncode(pass)='"+applicationEncode(Spass)+"','"

Observe that the encoding functions databaseEncode and application Encode are implemented at database layer and application layer, respectively.

2. **Transformation of traditional string operations into semantically equivalent operations in the new language**: We want our encoding function such that all the string operation in the encoded domain should provide same results as in the original domain. This can be achieved through homomorphism property. Homomorphic encoding allows specific types of computations on the encoded texts

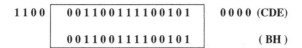

Fig. 2 Incorrect matching of encoded strings

such that the generated result matches with the result of the same operation on the original text. Observe that traditional substring matching operations may not satisfy homomorphic property in the proposed encoded domain, suppose we want to search substring BH in another string CDE. The encoded form is shown in Fig. 2. Note that, this encoding will result in an incorrect matching. To solve this using homomorphic property, we propose the following two possible enhancements:

(a) *Delimiter based encoding*: We now redefine the encoding function by introducing delimiter at the Boundary. For example,

$$F(CDE) = 01000011; 01000100; 01000101$$

In this example, the new alphabet has three symbols 0, 1, and;. As these three symbols cannot uniquely represented using single bit, the encoding function represents 0, 1, by 00, 11, and 010, respectively, as

$$F(CDE) = 0011000000001111010001100000011000001000110000001100011$$

(b) *Use predefined procedures at database layer*: We may also use any of the predefined procedure into database library such as Binary, Hex, etc. This ensures that matching is done on the byte boundary. For example,

"SELECT eid FROM loginfo WHERE BINARY(login)=" +BinEnc(Slogin)+ "' AND BINARY(pass) = '" +BinEnc(SPass)+ " ' "

or

"SELECT eid FROM loginfo WHERE HEX(login)=" +HexEnc(Slogin)+ "' AND HEX(pass) = '" +HexEnc(SPass)+ " ' "

where HEX and BINARY functions are predefine in database library. Hex converts strings into hexadecimal and BINARY convert strings into binary form. BinEnc and HexEnc are the encoded functions defined at application level.

Observe that the homomorphic property can easily be extended in case of other operations as well, e.g., LIKE, GROUP BY, etc.

4.4   Assertion-Based Approach

Instrumenting programs by adding assertions at critical positions of the programs is one of the most successful program verification approaches, which has various application domains, e.g., safety property checking, proving program correctness, automatic test-case generation and fuzzing, proof carrying code, input filter generation, etc. [5].

Program verification has recently received renewed attention from the Software Engineering community. One very general reason for this is the continuing and increasing pressure on industry to deliver software that can be certified as safe and correct. A more specific reason is that program verification methods fit very naturally the so-called design by contract methodology for software development, with the advent of program annotation languages like JML [13].

It is common for assertions to be defined as a super-set of Boolean expressions, since they may have to refer to the values of expressions in the current state of the program. If exactly the same syntax is used for assertions and Boolean expressions, it will be easier for ordinary programmers to write specifications.

Assertions that hold before and after execution of a program, preconditions and postconditions, respectively, will allow one to write specifications of programs or Hoare Triples. The intuitive meaning of a specification $\{P\}$ C $\{Q\}$ is that if the program C is executed in an initial state in which the assertion (precondition) P is true, then either execution of C does not terminate or, if it does, assertion Q (a postcondition) will be true in the final state. As example, $\{a > 3\}\ a = a + 7\ \{a > 10\}$ represents that, for any state satisfying $a > 3$, the execution of $a = a + 7$ will end with satisfying $a > 10$.

We extend the assertion based program verification to detect and prevent SQLIA. The approach consists of the following phases:

1. Preprocessing the input strings by filtering meta-characters.
2. Identify l_i: Q_i ($\{I_{ij} | j = 1, \ldots, m\}$)
3. Insert assertion after each l_i which checks the correctness of Q_i according to the specification.

Observe that assertion can be implemented either at database level or at application level. An application level assertion is illustrated in the following example: Consider a web application which contains the following:

```
rs = "SELECT eid FROM loginfo WHERE login='"+Slogin+"' AND pass='"+Spass+"' "
```

As bad inputs may change the semantics of the above query, our approach will add a piece of code (assertion) in the web application which checks the content of rs during run time in order to verify that no SQLIA happens. Below is a sample application-level assertion checking code:

Table 1 Comparison of proposed techniques w.r.t. existing ones in the literature

Technique	Auto-detection	Auto-prevention	Identification of all input sources	False positive	False negative	Modify code base	Time complexity
Java Static Tainting [15]	✓	✗	✓	✗	✓	✗	$O(n)$
Security Gateway [19]	✗	✓	✗	✗	✗	✗	$O(n)$
WebSSARI [18]	✓	✗	✓	✗	✗	✗	$O(n)$
SQLrand [2]	✓	✓	✗	✓	✓	✓	$O(n)$
SQL DOM [16]	N/A	✓	N/A	✗	✓	✓	$O(n)$
IDS [22]	✓	✗	✓	✗	✓	✗	$O(n)$
JDBC-Checker [7]	✓	✗	✗	✗	✗	✗	$O(n)$
AMNESIA [9]	✓	✓	✓	✓	✓	✗	$O(2n)$
CANDID [1]	✓	✓	✓	✓	✓	✗	$O(n)$
CIAOs [14]	✓	✓	✓	✗	✗	✗	$O(n)$
DIGLOSSIA [20]	✓	✓	✗	✓	✗	✗	$O(n)$
Query Rewriting	✓	✓	✓	✗	✗	✓	$O(n)$
Encoding-based	✓	✓	✓	✗	✗	✗	$O(n)$
Assertion-based	✓	✓	✓	✗	✗	✗	$O(nm)$

```
int flag = 0 ;
while (rs.next) {
    if (!Slogin.equals(rs.getString(1)) || !Spass.equals(rs.getString(2))) {
        flag = 1 ;
        break ;
    }
}
if (flag == 1)
    print (SQL Injection attack);
```

We may also implement the assertion-based checking at database level. Observed that this technique introduces much computations overhead in case of large database.

5 Discussion and Complexity Analysis

The complexity of Query Rewriting approach mainly depends on (i) parsing, (ii) data insertion, and (iii) Nested query execution on "Input" table. The worst-case time complexity of LALR parser is linear w.r.t. the number of inputs. Considering less number of inputs in practice, this approach does not introduce much overhead at all. The complexity of second approach depends on the encoding domain. The third approach introduces a high-overhead in computational complexity because of runtime data checking. Overall, query rewriting-based approach is suitable for web application containing simple queries with few inputs and few nested forms. The encoded-based technique is suitable for web application involving few operations on database data. The third approach is suitable for small database web applications. Below, in Table 1, we provide a comparative analysis of our proposed approaches w.r.t. the existing ones. In the last column, we denote by n the number of SQL queries in the application and by m the number of tuples in the database.

6 Conclusion

This paper proposes three new approaches to detect and prevent SQL Injection Attacks. The proposals can be treated as alternative solutions w.r.t. the existing ones in the literature. This paper also described the advantages and shortcoming of each proposed technique w.r.t. applicability point of view. We are now in process of implementing tools based on the proposals.

References

1. Bisht, P., Madhusudan, P., Venkatakrishnan, V.N.: Candid: dynamic candidate evaluations for automatic prevention of sql injection attacks. ACM Trans. Inf. Syst. Secur. **13**(2), 14:1–14:39 (2010)
2. Boyd, S.W., Keromytis, A.D.: Sqlrand: preventing sql injection attacks. In: In Proceedings of the 2nd ACNS Conference, pp. 292–302 (2004)
3. Buehrer, G., Weide, B.W., Sivilotti, P.A.G.: Using parse tree validation to prevent sql injection attacks. In: Proceedings of the 5th International Workshop on SEM, pp. 106–113. ACM (2005)
4. Clarke, J.: SQL Injection Attacks and Defense, 1st edn. Syngress Publishing, New York (2009)
5. Comini, M., Gori, R., Levi, G.: Assertion based inductive verification methods for logic programs. Electr. Notes Theor. Comput. Sci. **40**, 52–69 (2000)
6. Cook, W.R.: Safe query objects: statically typed objects as remotely executable queries. In: In Proceedings of the 27th ICSE, pp. 97–106. ACM (2005)
7. Gould, C., Su, Z., Devanbu, P.: Jdbc checker: a static analysis tool for sql/jdbc applications. In: Proceedings of the 26th ICSE, pp. 697–698. IEEE Computer Society (2004)
8. Halder, R., Cortesi, A.: Obfuscation-based analysis of SQL injection attacks. In: Proceedings of the 15th IEEE Symposium ISCC, pp. 931–938. IEEE (2010)
9. Halfond, W.G.J., Orso, A.: Amnesia: analysis and monitoring for neutralizing sql-injection attacks. In: Proceedings of the 20th IEEE/ACM ASE, pp. 174–183. ACM (2005)
10. Halfond, W.G.J., Orso, A.: Combining static analysis and runtime monitoring to counter sql-injection attacks. SIGSOFT Softw. Eng. Notes **30**(4), 1–7 (2005)
11. Halfond, W.G., Viegas, J., Orso, A.: A classification of SQL-injection attacks and countermeasures. In: Proceedings of the IEEE International Symposium on Secure Software Engineering. IEEE (2006)
12. Huang, Y.W., Yu, F., Hang, C., Tsai, C.H., Lee, D.T., Kuo, S.Y.: Securing web application code by static analysis and runtime protection. In: Proceedings of the 13th International Conference on WWW, pp. 40–52. ACM (2004)
13. Leavens, G.T., Baker, A.L., Ruby, C.: Preliminary design of jml: a behavioral interface specification language for java. SIGSOFT Softw. Eng. Notes **31**(3), 1–38 (2006)
14. Lin, J., Chen, J., Liu, C.: An automatic mechanism for adjusting validation function. In: 22nd AINA, 2008, pp. 602–607. IEEE Computer Society, Okinawa, Japan (2008)
15. Livshits, V.B., Lam, M.S.: Finding security vulnerabilities in java applications with static analysis. In: Proceedings of the 14th Conference on USENIX Security Symposium, vol. 14, p. 18. USENIX Association (2005)
16. Mcclure, R.A., Krger, I.H.: Sql dom: compile time checking of dynamic sql statements. In: ICSE05: Proceedings of the 27th ICSE, pp. 88–96. ACM (2005)
17. Nguyen-Tuong, A., Guarnieri, S., Greene, D., Shirley, J., Evans, D.: Automatically hardening web applications using precise tainting. In: Security and Privacy in the Age of Ubiquitous Computing, IFIP TC11 20th International Conference on SEC, pp. 295–308 (2005)
18. Ray, D., Ligatti, J.: Defining code-injection attacks. In: Proceedings of the 39th POPL, pp. 179–190. ACM (2012)
19. Scott, D., Sharp, R.: Abstracting application-level web security. In: Proceedings of the 11th International Conference on WWW, pp. 396–407. ACM (2002)
20. Son, S., McKinley, K.S., Shmatikov, V.: Diglossia: detecting code injection attacks with precision and efficiency. In: Proceedings of the 2013 ACM SIGSAC Conference on Computer & Communications Security, pp. 1181–1192. ACM (2013)

21. Su, Z., Wassermann, G.: The essence of command injection attacks in web applications. In: Conference Record of the 33rd POPL, pp. 372–382. ACM (2006)
22. Valeur, F., Mutz, D., Vigna, G.: A learning-based approach to the detection of sql attacks. In: Proceedings of the 2nd International Conference on Detection of Intrusions and Malware, and Vulnerability Assessment, pp. 123–140. Springer (2005)

Securing Service in Remote Healthcare

**Tapalina Bhattasali, Rituparna Chaki, Nabendu Chaki
and Khalid Saeed**

Abstract Health-care service in remote environment opens for several security challenges. These may affect confidentiality, integrity, and availability of resource. Securing service is a big concern for this kind of application. Encoding is required before uploading data to remote web server. Identity management is another primary aspect to validate any service. One-time identity verification during login has no importance, because valid session may be hijacked by impostors. Compared to other techniques, identity management based on human computer interaction is simple and less costly in remote environment. Service verification also needs to be considered to control access rights along with end user verification. A secured remote service (SecReS) framework is proposed here to ensure availability of health-care resource to valid end users. This service is capable to reduce time complexity, bandwidth cost, and to increase accuracy and attack resistance capacity. Theoretical analysis shows its efficiency.

Keywords Data retrieval · Privacy preservation · Remote healthcare · End user verification · Service verification · Identity management · Hybrid encoding

T. Bhattasali (✉) · R. Chaki · N. Chaki
University of Calcutta, Kolkata, India
e-mail: tapolinab@gmail.com

R. Chaki
e-mail: rituchaki@gmail.com

N. Chaki
e-mail: nchaki@gmail.com

K. Saeed
Bialystok University of Technology, Bialystok, Poland
e-mail: khalids@wp.pl

© Springer India 2016
R. Chaki et al. (eds.), *Advanced Computing and Systems for Security*,
Advances in Intelligent Systems and Computing 395,
DOI 10.1007/978-81-322-2650-5_5

1 Introduction

In remote health-care application, relevant data are collected and stored into remote storage; so that it can be accessed from anywhere and at any time for medical check-up. Health-care professionals from remote location provide medical assistance based on the accessed information. Pervasive nature of remote health faces several known and unknown security vulnerabilities [1, 2], which lead the entire framework unreliable, putting patients' lives at risk. Lack of proper security framework is one of the major barriers in long-term success of remote health application. Major security requirement for remote environment is to restrict unauthorized access to resource [3]. It must be ensured that openness is provided only to the users who are authorized to access. Remote health framework provides additional layer of security with this approach, thereby reducing risk of accessing health-related data by unwanted users [4].

Popularity of identity management through analyzing human behavior is growing day by day. Behavioral traits are capable to provide better result than traditional identity verification systems based on passwords or tokens [5]. As human interaction pattern [6] is naturally related with manner in which individual interacts with computing device, it is difficult to be compromised. Most of the behavioral traits require data acquisition devices separately, which may not be always available in remote areas or which may enhance the cost of the applications [7]. Service verification used along with the end user verification provides equivalent features of role-based access control. This concept is feasible for any service that needs long-term access from remote places.

Traditional data retrieval based on plain text becomes useless in this regard [8]. Uploaded cipher data need to be decoded before its usage. If decryption key revealed to unknown users, then privacy of data is compromised. The best way of securing health-care resource from unauthorized persons is identity verification along with encryption. This paper considers three directions—(i) data retrieval service, (ii) identity management service including end user verification (EUV), and service verification (SV) and (iii) encoding service.

The rest of the paper is organized as follows. Section 2 presents background of proposed work. Section 3 describes proposed work. Section 4 includes analysis part followed by conclusion in Sect. 5.

2 Background of Proposed Work

2.1 Literature Survey

Researchers are very much interested to work on securing services in remote health. One of the important aspects of securing remote service is to consider data retrieval service. Most of the existing data retrieval logics are based on only Boolean keyword search [9, 10]. Search flexibility can be enhanced by considering more than one keyword in a single query. Data are retrieved based on whether a searched item is present or not, without considering the degree of relevance [11]. Order preserving encryption (OPE) [12] is capable to search only one item in the query without considering privacy preservation. There also exist a few works on multiple search items [13, 14] in a single query. Ranked search enhances usability by returning the required data according to relevance rank [15, 16]. Topmost k data are retrieved including privacy preservation, secure indexing, and ranking. The problem of top-k [17] multiple search items retrieval over encrypted data is solved in modified MRSE (multi-keyword ranked search over encrypted cloud data) [18], which uses coordinate matching and inner product similarity. Query keywords [19] can be protected by using cryptographic Trapdoor [20].

Research efforts of securing service are also directed toward identity management, which is very useful in remote environment to identify validity of remote users [21, 22]. In remote framework, valid users can directly access data regularly from remote storage after giving biometric authentication proof. In this framework, local client is responsible for capturing user's behavioral data and sending it in encoded format to the web server, where the matching process is executed. Several techniques exist for securing biometric authentication [23]. Useful data are kept in trustworthy remote servers to reduce leakage. Sometimes users try to login to remote server by using smart card along with personal biometrics for verification [24–26]. Biometric sample is checked with smart card value. There exists three-factor authentication [27], where password, biometrics, and smart card are fused. To improve security and strong mutual authentication between user and remote server, long pseudo-random numbers and timestamps can be used.

Besides verifying valid users, the level of access [28] should also be defined. Otherwise, any user can access any resource after giving valid authentication proof; but this is not desirable in case of remote health environment. As for example, doctor and insurance agent must not have same level of access rights. Therefore, access right of end user must be predefined. However, access control mechanism alone has no use in remote framework. It needs to be considered along with authentication in remote environment. In HL7 RBAC [29] model, users are assigned access rights according to their static job roles.

After analyzing several existing works, it can be said that proof of retrievability [30], identity management and encoding play major roles in securing remote health service. This service should be designed in such a way that it provides flexibility as well as better results from the aspects of security and privacy. End Users' traits and sensitive data are kept secret from the web server to enhance privacy.

2.2 Compatibility Checking with Major Standards of Healthcare

Design of security solution must be compatible with two common standard regulations in remote healthcare, such as HIPAA [31], HL7 RBAC model [29], and PASS security model [32]. Table 1 represents possible security solutions compatible with standard regulations to avoid common risks in remote healthcare.

2.3 Scope of Work

Preserving privacy and effective use of encrypted resource are very important for designing secured remote health service. However, it is very much complex and difficult to control. Overall procedure must not be deterministic in nature. Web server must not be able to determine any relationship among the procedural logics. Accuracy level of pattern matching for original values and query values should be high. Response time of any service must be faster and query result must be more relevant. Encoding techniques used for privacy preservation need to have low time and compatibility with standard protocol used for secure data transmission in remote framework. Main aim of the proposed work is to consider effective identity management along with hybrid encoding during designing remote health service framework in a secured manner.

To activate secured remote service, it is expected to give the following security and performance requirements.

- Query contains multiple items and provides result based on highly relevant data to provide effective data retrieval service.
- Data retrieval service ensures fast response, low communication, and computation overhead along with privacy protection.
- Identity management is flexible enough to reduce time complexity, possibility of intrusion, and enhance accuracy level.
- Flexibility of HL7 RBAC Model is enhanced by proposed service verification.

3 Proposed Work

Proposed work of this paper is to design secured remote service framework (SecReS) that can provide quality health assistance to the patients [33]. In SecReS, patients need to give valid identity proof before uploading data to web server. Similarly, health-care practitioners or any other end users need to be validated before accessing data from web server. Access to data is blocked in the middle of the session if any unusual event is noticed. Third-party trust relationship [34, 35] is

Table 1 Secured remote health service compatible with standard regulations

Risks	Risk management solution in std. regulation	Limitations (if any)	Possible solution
For remote access			
Only user-ID and password are not sufficient; Security questions are difficult to remember	**HIPAA**-Two factor authentication; Use of RADIUS protocol	RADIUS protocol depends on unreliable transport protocol UDP and suffers from lack of security for large payload. MD5 algorithm used in this protocol is also proved as insecure	End user's behavioral pattern analysis along with password verification and service verification
Unauthorized user access sensitive data	**HL7 RBAC**-Establish role-based access control (RBAC)	Lack of flexibility	RBAC is modified by using service verification, where role can be changed dynamically according to context and time
Workstation left unattended, risking improper access	**HIPAA**-Establish appropriate solution for session termination (time-out)	It may result into false detection	Multi-time identity management to prevent impostors to enter into valid session
For storage and transmission			
Data access device is lost or stolen resulting into unauthorized access of data	**HIPAA**-Use of biometrics; Require use of lock down mechanism; Employ encryption technique of appropriate strength	All types of Biometric data acquisition devices may not be available anywhere and may increase cost	Behavioral biometrics based HCI improves performance; Semi-continuous mode verifies end user multiple times; Procedure becomes secure and fast with hybrid encoding
Sensitive data are left on external device	Prevent download of sensitive data to any device		Data retrieval is considered as a service
Data intercepted or modified during transmission or storage	Prohibit transmission of sensitive data via open network; Use of secure socket layer (SSL) protocol; Implement strong encryption solution		Use of existing virtual private network; Use of comparatively better transport layer security protocol (TLS); Low overhead and efficient cryptography solution with secure key management

considered here to reduce the probability of intrusion. Definition of proposed SecRes framework is given below.

Definition (SecReS) A secured remote service is a set containing five tuples i.e. service = {identity_check, encode, upload, download, decode}, entity_type is a two tuple set, i.e., entity_set = {end user, device} and member is a four tuple set, i.e., member = {DO, WC, WS, DU}, where data owner (DO) ∈ end user (EU), web client (WC) ∈ device (D), web server (WS) ∈ device (D), and data user (DU) ∈ end user (EU).

Members of the framework are defined below.

- **Data Owner (DO)**—End User is a member, who generates health data having {read, write, execute, upload, download} permission according to assigned service role. If service role = {patients ∨ local caregivers} then, entity ∈ DO. DO needs to give valid identity proof. Authentic DO records health data and uploads to web server.
- **Web Client (WC)**—Device is used by DO and DU and considered as interface for interaction with WS. WC generates and stores data temporarily. Data are encoded before uploading. During data transmission, end user profiles are collected and stored as template along with service profile and stored at WS side in encoded form.
- **Web Server (WS)**—Device has huge storage space and computational resources to maintain high volume of encoded data and templates uploaded from DO side. Decision is taken at WS side. DO can upload data and DU can access data from WS after giving effective identity proof.
- **Data User (DU)**—End User is a member, who can access health status of patients after giving valid identity proof. DU needs to decrypt data before using it. Default permission of authentic DU is {read, execute, download}, which can be modified according to requirement. If service role = {health-care practitioners ∨ patients ∨ relatives} then, entity ∈ DU. Health-care practitioners = {Doctors, nurses, caregivers, pathologists, health-care experts, insurance agents}.

Figure 1 represents flow of interaction of SecReS framework.

3.1 Procedure of Securing Major Service in Remote Healthcare

Definition (Secured Retrieval Service) A type of service in remote healthcare includes two more services-identity management service and encoding service. This service uses collection of following logics to retrieve health-care resource from remote server.

- construct_index()
- build_search_index()

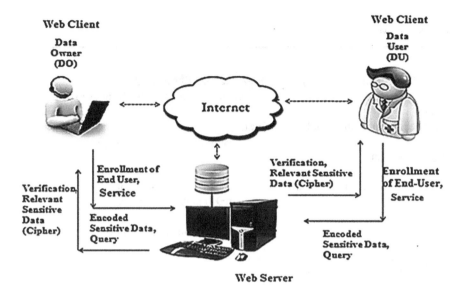

Fig. 1 Interaction flow of SecReS framework

- build_vec()
- encode()
- hybrid_encode()
- identity_check()
- gen_seed()
- gen_query()
- match_pattern()
- gen_rank()
- gen_reply()
- decode_data()

Procedural logic of SecReS framework is presented in Table 2.

3.2 Identity Management Service in Remote Healthcare

Definition (Identity Management) It is a four tuple service model {EUV, SV, timestamp, interval}, where timestamp represents session time and interval represents a nondeterministic function. Here end user verification EUV = {HIA, encoding, trust, add_info}, service verification SV = {access right, job role, duration, context}. Here human interaction analysis HIA = {fixed pattern, free pattern, temporal feature, global feature} and add_info = {device-id, operating system, locality, network-id, browser type, cookies}. Fixed pattern represents static

Table 2 Procedural logic of
SecReS framework

```
begin
    call construct_index(hd_p )              // input: sensitive data hd_p
                                             // output: encoded searchable index in_c
begin
in_p:=build_search_index (hd_p, stemming)
                                             // using stemming technique[36]
in_c:=encode (in_p)
end
call build_vec (m, in_c, tf, idf, n)         // m→no. of searchable items,
                                             // n→ number of records
                                             // tf→ term frequency,
                                             //id → inter-domain frequency

begin
for each hd_j
vec_j(m+1):=build_vec(tf,idf)[36]
end for
si_n:={vec_j|1 ≤ i ≤ n}                       // serachable index
end
call hybrid_encode(aes, ecc)
call md(sha, hd)                             // message digest
call identity_check (euv, sv)
seed_val:= gen_seed(query q)
call secure_seed(seed_val)
call gen_query (vs_model, query q)  [37]
                                             // vs_model →vector space model
nn_score:= match_pattern(deg_of_similarity)
call gen_rank(nn_score)
call gen_reply( rank_id, reply_vec, top_j)
call decode_data(aes, ecc)

end
```

---

interaction, whereas free pattern represents flexible interaction. Nontemporal data are represented as global feature.

Validity decision of identity management mainly depends on {EUV∧ SV} for multiple times. Figure 2 represents identity management concept.

End User Verification (EUV).

Definition (EUV) In EUV, end users are verified by two factors {password, interaction analysis}, i.e., password itself and human computer interaction analysis (IA) of password as well as any interaction by user. Any frequently used pattern is

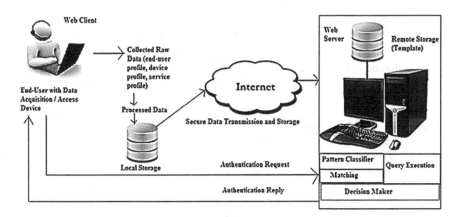

Fig. 2 Identity management concept in SecReS framework

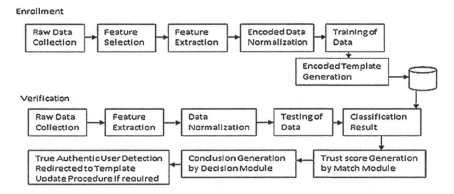

Fig. 3 Proposed end user verification

represented by $\sum$. Temporal data are represented in discrete form, where $T = \{t1, t2, \ldots, tn\}$. User generated patterns follow a sequence. IA is defined as 4-tuple $\{t, itime, a, n\}$, where starting timestamp of session $t \in T$, interaction time itime $\in T$, action $a \in \sum$, n-graph $n \in \sum$. Function fs determines whether interaction pattern is part of interaction sequence S, where fs: $t \times t \times \sum \times \sum \rightarrow \{0,1\}$.

Figure 3 presents basic overview of proposed end user verification. The major steps of interaction analysis are given below. It captures user's usage pattern at random interval.

Raw Data Collection Samples are collected from web-based environment. Client-side application collects raw data. n-graph is considered as size of atomic interaction.

Feature Selection Parameters such as user-id, session time, interaction time, and actions are selected for processing feature vector. Other collected parameters are stored temporarily in local storage and transmitted toward web server after

Table 3 Filtering logic

---

if itime (n_i) < itime $|m_n - \sigma_n|$ itime or if itime (ni) > itime $|m_n + \sigma_n|$ then,
   discard itime (n_i)

                          // where m_n is median of interaction time
                             for all type of n-graphs and σ_n is standard
                             deviation of interaction time for all n-graphs
if itime(n_i) < itime $(m_n,s_j - \sigma_n,s_j \times 2)$ or
   itime(n_i) > itime $(m_n,s_j + \sigma_n,s_j \times 2)$ then,
      discard itime(n_i)

                           // where m_n,s_j median of interaction times of
                              all occurrences of that type of n-graph
                              in session s_j and σ_n,s_j is standard deviation of
                              interaction times of all n-graphs in session s_j

---

encoding. They are considered as input for query execution during EUV, the result of which is taken as input of decision maker.

Filtering If interaction times for specific type of n-graphs are too short then, it may result into errors in the measurement. If interaction times of the n-graphs are too long then, it may be due to hesitation in user's behavior. These interaction times are not representative of user's interactive behavior and need to be discarded. Filtering of interaction times are given in Table 3.

Feature Extraction Selected features are extracted and mainly processed based on interaction times. Different feature vectors (FV) are constructed as follows.

FV1 includes {uid, from interaction code, to interaction code, avg. interaction time}

FV2 includes {uid, pattern_size, mean interaction, median interaction, sd interaction}, where sd $\rightarrow$ standard deviation

FV3 includes {uid, time per_unit_action, overall speed}

Template Generation Template stores normalized temporal data for each user that enables interaction analysis to distinguish between valid user and an invalid user. Frequently used patterns are considered during classification. Feature vectors are stored at local storage as template according to priority. Here FV1 is assigned priority 1, FV2 is assigned priority 2 and FV3 is assigned priority 3. Remaining feature vectors are assigned priority number onwards. They are encoded and transmitted towards web server in a secure way (using TLS protocol). During verification, feature vectors are searched from the templates according to the assigned priority values.

Data Classification Normalized template data are trained offline to improve performance of testing required during verification phase. Classification result has major impact on accuracy of verification. Single-class classification is designed here

Table 4 Data classification
logic

---

query (q) to classifier ← claimed user' s verfication features
assign non-binary weight to fv_i for both template(t)and query(q)
associate weight $w_{i,j}$ with a pair $\{fv_i, t_j\}$,
    where $w_{i,j} >= 0$ and non-binary and fv_i is feature vector and t_j is template
                                                  vector
associate weight wi,q with a pair $\{fv_i, q\}$,
    where $w_{i,q} >= 0$ and non-binary and fv_i is feature vector and q is query
                                                  vector
query vector $q_{vec} = \{w_{1,q}, w_{2,q}, .., w_{n,q}\}$,
template vector $t_{vec} = \{t_{1,j}, t_{2,j}, ..., t_{n,q}\}$
                        where n is the total number of features
$sim(t_{vec}, q) := (t_{vec}. q_{vec} / |t_{vec}|.|q_{vec}|)$
                            // correlation between t_{vec} and q
$$sim(t_{vec}, q_{vec}) = \frac{\sum_{i=1}^{n} m \times r}{\sum_{i=1}^{n} m^2 \times \sum_{i=1}^{m} r^2}$$ where $m = w_{i,j}$ and $r = w_{i,q}$

dos ← degree of similarity sim()
sort dos in descending order
evaluate_rank (dos)

Table 5 Match logic

---

if rank ranges between 0 to +1 then,
    match_score = 1
else
    match_score= 0

---

for open setting environment. Data Classification is obtained by using vector space
model logic. Vector space model is most popular one in information for matching
pattern. Data classification logic is presented in Table 4.

Match Score Generation Match module generates match score based on classification result. Match logic is presented in Table 5.

   add_info() of EUV includes used device information, which is an important step
in uncontrolled environment. Devices used by end users are registered during its
first usage. This concept is almost similar to device fingerprinting [38]. However,
the complexity and hardware dependency are reduced in this approach.

Service Verification (SV).

The basic concept is based on role-based access control (RBAC) with great flexibility and robustness. The major advantage of service verification is its ability to make access control decisions dynamically based on the context information. This property is mainly useful for applications like remote healthcare. It avoids the limitation of static RBAC as proposed by HL7 RBAC model [29].

Definition (SV) A type of identity management is defined as a quadruple, i.e., SV = {ar, role, dur, cntxt}, where ar represents access-right, role represents job role, dur represents duration of session, cntxt represents context.

Relationships among tuples of SA are as follows.

ua(User Assignment) ϵ uid $\times$ role $\rightarrow$ many to many relation for users to set of roles mapping

one user is assigned to set of roles $\rightarrow$ one to many mapping

ar(Access Right) ϵ AR $\times$ role $\rightarrow$ many to many relation for access right to role mapping

mapping of a role onto a set of permissions $\rightarrow$ one to many mapping

mapping of user onto a set of durations (dur) $\rightarrow$ one to many mapping

mapping of durations onto set of roles $\rightarrow$ one to many relationship.

In SV, each end user is assigned a role, which is subset of entire set of job role, each job role is assigned access right, which is subset of entire set of access right set. To design SV dynamically, state machines are considered to maintain role for each user and access right for each role. State transition occurs when contexts are changed.

If access right = null, end user has no right to interact with WS.

State Transition is defined as ST(SS, DS), where SS $\rightarrow$ source state, DS $\rightarrow$ destination state. As for example, consider there are two access rights {ar1, ar2}. As for example, ar1 implies read and write permission, while ar2 implies only read permission. ar1 is assigned when system load is low, but transition occurs from ar1 to ar2, when system load is high. The major steps of service verification are given below.

Data Collection Raw data are collected once during enrollment of end users besides collecting interaction data. Normally, this type of collection occurs once. If any of the tuples of SV need to be changed according to the requirements of environment, only specific tuples are modified. The tuples of SV are defined as follows.

Access Right (ar)—It represents approval to access one or more protected resource. It is a set of permissions such as {read, write, execute, upload, download}.

Job role (role)—It is a job function assigned according to the context of the environment regarding the responsibility. It represents a set of roles. It can be changed dynamically according to the requirements.

Duration (dur)—It represent show long a user is assigned a set of roles. Duration is a set of one or more sessions when user interacts with resource.

Table 6 Evaluation of service verification match score

---

if SAV matching ranges between 0.9 to 1then,
    match_score:=1
else
    match_score:=0

Context (cntxt)—It represents the set of context information around the environment. This parameter is very important in dynamic environment. It enhances the flexibility of the approach.

Feature Extraction and storage Selected features are extracted and feature vector is constructed.

Service Verification Vector (SVV) includes {uid, ar, role, dur, cntxt}
This feature vector is temporarily stored at local storage, and then it is encoded and transmitted securely toward web server to store the encoded data in a secure way. During verification, optimized query is generated to match service profile of claimed end user's identity. As for example, when claimed user demands for a service, then service profile of that user needs to be verified to check whether that user has permission for that service. If service profile is not matched, then alert is generated. Service verification match score evaluation is presented in Table 6.

Fusion of Multiple Factors

Decision Generation Finally, decision module takes the decision of identity management based on fusion of match scores, generated by two factors—EUV, SV. Fusion logic is presented in Table 7.

3.3 Procedure of Hybrid Encoding Logic in Remote Healthcare

Encoding is an important security feature to ensure confidentiality and integrity in remote framework. Encryption protects relevant sensitive data and secret template data from unauthorized users. Integrity can be verified by using digital signature. As framework deals with sensitive health data of patients, confidentiality, and integrity of data required for verification must be ensured. Existing symmetric encryption schemes are capable to provide high security, but key maintenance is a considerable issue. In asymmetric encoding, key management becomes easier. However, complexity of the encoding enhances with a possibility of lack of confidentiality or integrity. Security of symmetric encoding is high because of presence of large

Table 7 Fusion logic

confidence := $(w_1 \times s_1 + w_2 \times s_2)/n$
 where n represents number of factors involved in identity management
 where $s_1 \rightarrow$ match score generated by EUV and $w_1 \rightarrow$ weight assigned to
 EUV according to its priority in multifactor identification.
set w_1 to 2 (high priority)
 where $s_2 \rightarrow$ match score generated by SV
 and $w_2 \rightarrow$ weight assigned to SV according
 to its priority in multifactor identification
 set w_2 to 1
if confidence >=1 or confidence <=2 then,
 end-user is valid
else
 end-user is invalid

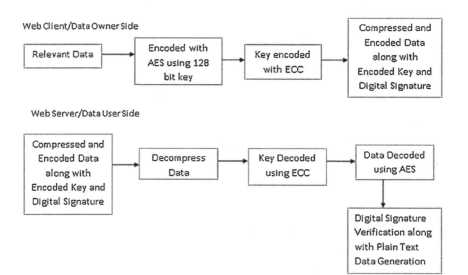

Fig. 4 Hybrid cryptography to secure resource

number of rounds. Advanced encryption standard (AES) is faster and less intrusive. Elliptic curve cryptography (ECC) is considered as feasible asymmetric encoding in heterogeneous domain because of its low resource consumption. For this reason, hybrid encoding logic is considered here by combining AES and ECC for enhanced security. AES key is encoded with ECC to avoid key compromise. Figure 4 represents basic idea of hybrid cryptography.

4 Analysis

This section contains the analysis part to validate proposed secured remote service (SecReS) framework. For providing real-time service, time is a major concern. Here, hybrid cryptography logic applied on relevant data (medical data, template). Sample medical data are encoded by AES (Rijndeal) block cipher and analyzed by CrypTool 1.4.31 Beta 6b.

Simple key value considered here is 00000000000000000000000000000000.

Sample Medical data in plain text is presented in Fig. 5.

Sample medical data in cipher text is presented in Fig. 6.

Hash value of the above record at data owner side is 44 15 38 0E B0 BC D1 49 B3 F2 3F CC 85 E4 5E F6 C5 F1 4B 8A

Hash value of the same record at data user side is 44 15 38 0E B0 BC D1 49 B3 F2 3F CC 85 E4 5E F6 C5 F1 4B 8A

Difference between hash values of the record at sender and receiver side is given below.

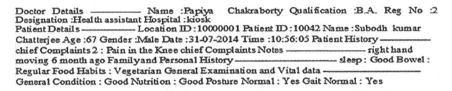

Doctor Details ———— Name :Papiya Chakraborty Qualification :B.A. Reg No :2 Designation :Health assistant Hospital :kiosk
Patient Details———— Location ID : 10000001 Patient ID : 10042 Name :Subodh kumar Chatterjee Age :67 Gender :Male Date :31-07-2014 Time :10:56:05 Patient History———— chief Complaints 2 : Pain in the Knee chief Complaints Notes ———————— right hand moving 6 month ago Family and Personal History———————— sleep : Good Bowel : Regular Food Habits : Vegetarian General Examination and Vital data ———————— General Condition : Good Nutrition : Good Posture Normal : Yes Gait Normal : Yes

Fig. 5 Sample medical record (plain text)

```
98 CB 46 26 12 21 C6 8E 37 58 FC 45 38 C3 93 DC 07 DF 5F CA 9A 43 B0 45 21 9A DE 92 17 29 8B D6
7A 91 12 4B 21 BD 52 ES A7 28 D4 6D 59 4E FC 18 8D E B 4D 53 02 A6 02 54 00 A7 D0 FA FF 37 85 A3
73 68 B9 C9 1C 88 A9 B6 4E B1 4E CF A9 84 A4 4F 88 C2 21 33 A9 65 3A 48 3B F7 49 E0 52 E4 E6 49
85 76 30 60 37 05 F9 5B 08 73 A0 0F AB 54 B2 E2 76 C2 74 93 1C 45 13 A3 00 1A 44 0C 54 9D 8D E4
42 F2 C6 F2 A6 BA 35 2F FB BB 01 4C 3A 1C A7 36 E6 D8 52 4F 56 9A CF 60 E6 17 4F 2A 78 1D F3 F0
E1 24 06 E9 D9 40 B0 0B CF AD 7C BB 42 DC AF 39 2C 7C 8A 55 5D 5A 00 5C F7 BD A8 29 63 C1 1D E9
CE 89 76 0D AA 3B D8 1D 16 D2 1C B8 78 6B 4E C3 F0 F6 A6 07 5B 9E E9 FC 84 32 C1 B0 D8 88 05 0F
85 9C 1A 6C D6 0B 55 55 65 6F 75 08 E8 08 7D 26 B3 87 03 8D 14 00 F1 65 93 4A 44 BC 0A 28 6E F5
7F C9 04 C1 74 D8 A5 46 29 3F 55 10 E3 27 63 DB 1A 59 DC 28 CS E0 17 A4 1D B4 AD 30 19 9E EA 24
7D A0 44 8C 00 3C 2D D6 9F 04 BA E4 07 64 93 98 1A 91 78 70 50 CS AA 3A D1 5B 6E A8 4A 2E DA BF
33 6A 9C 5A 69 49 7E 59 98 18 E7 A4 11 92 FA 20 97 5C 5D B8 87 51 25 77 5A 11 CC BC 4D F3 AC F7
41 3B 14 3A 2B C2 22 80 4B D4 CA 41 42 40 7B 66 D5 86 45 78 CB 0C E9 F5 FC 93 88 65 37 F6 CF 37
F6 E9 70 3A 4B 19 E F 80 5F 84 80 A0 69 7F 84 B4 C9 92 CE EA 2A 02 5F 5E 22 57 B7 2C 62 5E 7D 50
00 CS 35 47 3D 80 11 5F 4F 52 54 D0 0E 4A 16 51 80 99 AB 92 74 4C 99 83 29 E2 8C BC 89 67 89 AD
77 BF 4F 2D EB 11 BA E1 21 A6 B8 B1 0E 6E B3 52 46 9A 8D 55 1F 78 AE A8 6C 98 47 47 EB 3A AA 9A
CD 48 A8 8C 6C 74 78 45 9E E4 7B 2B 5F 1D E5 3D C6 D8 E5 4A CF 0F 2D 05 44 8C 3D 29 D7 59 ES AC
1B 98 AD 2A 43 1C 38 D1 C5 60 8C 00 AA F6 C5 55 E5 78 8A 68 32 AD 3C A7 D3 4D C6 03 9F 92 A0 B2
82 F1 9A DC 46 CB C3 64 1A FA 87 D7 EE F9 61 85 30 CS AE 73 09 9F FC DC 00 A6 86 E B 09 5D 3C 04
56 2C B7 67 74 E4 E9 B3 3F BB A1 86 77 3A 2F 33 FF 26 CE 75 AE 24 19 7B 9B 6B 46 80 37 B0 33 EA
5F B2 6F 5A 66 56 80 80 06 4A 2F 5C 79 8F 22 A6 71 CF E0 8F E3 6B 10 04 AA D3 7B 5E 06 E7 2 CF0
0B E6 5B F5 AE 9F E6 AE CB 99 C7 82 32 0D 26 36 22 76 A3 CA EE 49 97 D9 E6 F4 FC C3 17 03 16 27
37 F9 5E 62 E B DA 78 C2 A2 D4 85 29 67 84 95 7B 3B 14 3B 72 03 69 50 B2 23 AC 72 C3 A5 5D 5 D37
1E 24 C1 B0 19 8A 1B 6F 20 C6 A4 57 ES 5E B1 CF B0 D2 24 6D 05 C7 65 0D A7 6E 86 F3 07 79 17 D8
```

Fig. 6 Sample medical record (cipher text)

00000000#00000000#00000000#00000000#00000000#00000000#00000000#0
0000000#00000000#00000000#00000000#00000000#00000000#00000000#0000
0000#00000000#00000000#00000000#00000000#00000000#

0.00 % of the bits differ (0 of 160).

Longest identical bit sequence: offset 0, length 160.

After analyzing domain parameters to be used for used key EC-prime239v1 with PIN 1234, it is seen that digital signature consists of two numbers c and d given below. The algorithm used here is ECSP-DSA with hash function SHA-1(160 bits).

Secret interaction s of signature originator =
2011757352490501151490225574977387280663081106137622911739360039
32280863

Hash value f (message representative) from message M, using the chosen hash function SHA-1,

f = 680948017289933777203873644291425651407543321486

c = 30456662765821009506612572133651168145735012845318676248592334227879747

d = 7038158704102622667665036483399955162595378135159329968256205

14433900631

Signature length is 474 bits. Signature generation time is 0.004 s. Correct signature verification time is 0.010 s. Entropy of the signature is 3.76. ECC encrypted AES session key length is 121 byte. Encryption time of above plain text data is 0.149 s; compression rate of cipher text is 10 %. Decryption time is 0.082 s.

The sample record (AES cipher) in Fig. 6 contains 240 different byte values. Entropy of the whole record is 7.71. High entropy value and less correlation among the byte values of cipher reduce the possibility of attack. Figure 7 represents comparative analysis of entropy and different byte values of commonly used symmetric cipher. It is seen that AES symmetric cipher shows less deterministic nature, which reduces the probability of data compromise.

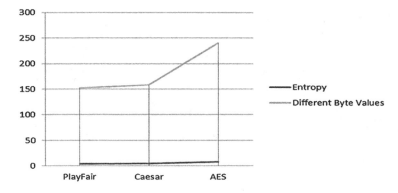

Fig. 7 Entropy versus different byte values for few symmetric ciphers

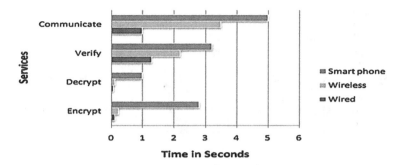

Fig. 8 Plot of delay for different devices

Identification Delay (ID) is the time gap between the instance when claimed user's data taken as input for verification and the instance when identification decision is taken.

ID = Encryption time + Decryption time + Verification time + Communication time

Figure 8 gives the idea about delay for different ubiquitous devices. It is seen that identification delay becomes higher in smart phone like resource constraint devices and becomes lower in wired devices.

Figure 9 represents session hijacking probability versus resource consumption for different modes of identity management. It is seen that session hijacking probability is high in one-time verification with low resource consumption, whereas session hijacking probability is less in continuous verification with high resource consumption rate. Therefore, identity verification for multiple times is feasible one.

Figure 10 shows accuracy rate for different types of identity management at different sessions. Accuracy is proportional to number of factors involved in identity management logic. Therefore, it is seen that proposed multifactor (EUV + SV) logic is capable to provide high accuracy.

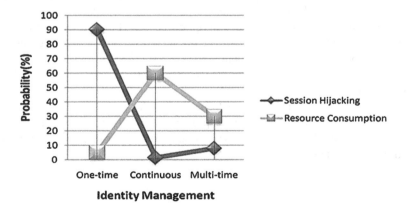

Fig. 9 Plot of session hijacking probability (%) versus resource consumption (%)

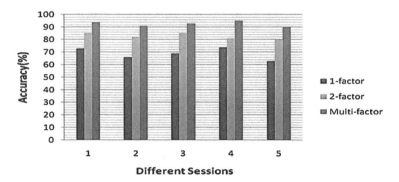

Fig. 10 Accuracy (%) for different types of identity management

This analysis part ensures that proposed secured service shows reliable result in remote health framework in terms of security vulnerabilities, speed, and accuracy.

5 Conclusion

In this paper, a procedural logic is proposed to provide secured remote health service. This logic supports CIA property of security. Major requirement to provide secured service in remote healthcare is to consider identity verification and encoding. Proposed logic focuses on secured data retrieval service that includes identity management and hybrid encoding. Use of vector space model and calculation of nearest neighbor score based on tf × idf weight value enhance the efficiency of service based on rank. Among various biometrics, human interaction analysis is proved to be more simple and cost-effective. Device enrollment as an additional feature of end user verification reduces the complexity and hardware dependency of device (radio) fingerprinting. Service verification improves the flexibility of role-based access control based on time and context. Besides executing pattern matching, queries are executed in optimized way. Finally, decision module fuses match scores from all factors to take the final decision. Instead of executing verification logic one time or in a continuous way, proposed logic executes multiple times. AES block cipher is considered to encode relevant data and template. ECC is used to secure AES.

At present, work is on for detailed analysis of the proposed logic in remote health-care domain. The main aim is to provide better service in a secure way in health-care domain. In future, this service is planned to be extended as a service in cloud on pay per use basis.

References

1. Aleman, J.L.F., Senor, I.C., Lozoya, P.A., Toval, A.: Security and privacy in electronic health records: A systematic literature review. J. Biomed. Inf. **46**(3) 541–562 (2013)
2. Bhattasali, T., Chaki, R., Chaki, N.: Study of security issues in pervasive environment of next generation internet of things. In: Proceedings of Computer Information Systems and Industrial Management (CISIM). Lecture Notes in Computer Science (LNCS), vol. 8104, pp. 206–217. Springer, New York (2013)
3. Belsis, P., Pantziou, G.: A k-anonymity privacy-preserving approach in wireless medical monitoring environments. Pers. Ubiquit. Comput. **18**(1), 11–74 (2014)
4. Bhattasali, T., Saeed, K.: Two factor remote authentication in healthcare. In: Proceedings of IEEE International Conference on Advances in Computing, Communications and Informatics, pp. 380–381 (2014)
5. Barkhuus, L.: The mismeasurement of privacy: using contextual integrity to reconsider privacy in HCI. In: Proceedings of the SIGCHI Conference on Human Factors in Computing Systems, pp. 367–376 (2012)
6. Castro, L.A., Favela, J., Quintana, E., Perez, M.: Behavioral data gathering for assessing functional status and health in older adults using mobile phones. Pers. Ubiquitous Comput. **19**(2), 379–391 (2015)
7. Bhattasali, T., Saeed, K., Chaki, N., Chaki, R.: A survey of security and privacy issues for biometrics based remote authentication in cloud. In: Proceedings of Computer Information Systems and Industrial Management (CISIM). Lecture Notes in Computer Science (LNCS), vol. 8838, pp. 112–121. Springer, New York (2014)
8. Boneh, D., Crescenzo, G., Ostrovsky, R., Persiano,G.: Public key encryption with keyword search. In: Proceedings of International Conference Theory and Applications of Cryptographic Techniques (EUROCRYPT). Lecture Notes in Computer Science (LNCS), vol. 3027, pp. 506–522, Springer, New York (2004)
9. Curtmola, R., Garay, J. A, Kamara, S., Ostrovsky, R.: Searchable symmetric encryption: improved definitions and efficient constructions. In: Proceedings of ACM Conference Computer and Communication Security (CCS), pp. 79–88 (2006)
10. Wang, C., Cao, N., Li, J., Ren, K., Lou, W. : Secure ranked keyword search over encrypted cloud data. In: Proceedings of IEEE International Conference Distributed Computing Systems (ICDCS), pp. 253–262 (2010)
11. Swaminathan, A., Mao, Y., Su, G.M., Gou, H., Varna, A.L., He, S., Wu, M., Oard, D.W. : Confidentiality-preserving rank-ordered search. In: Proceedings of Workshop of Storage Security and Survivability, pp. 7–12 (2007)
12. Martinez, S., Miret, J.M., Tomas, R., Valls, M.: Security analysis of order preserving symmetric cryptography. Appl. Math. Inf. Sci. **7**(4), 1285–1295 (2013)
13. Cao, N., Wang, C., Li, M., Ren, K., Lou, W.: Privacy-preserving multi-keyword ranked search over encrypted cloud data. IEEE Trans. Parallel Distrib. Syst. **25**(1), 222–233 (2014)
14. Ming, L., Yu, S., Cao, N., Lou, W.: Authorized private keyword search over encrypted data in cloud computing. In: Proceedings of IEEE International conference on distributed computing systems, pp. 383–392 (2011)
15. Wang, C., Cao, N., Ren, K., Lou, W.: Enabling secure and efficient ranked keyword search over outsourced cloud data. IEEE Trans. Parallel Distrib. Syst. **23**(8), 1467–1479 (2012)
16. Ming, L., Yu, S., Cao, N., Lou, W.: Toward privacy-assured and searchable cloud data storage services. IEEE Trans. Netw. **27**(4), 56–62 (2013)
17. Yu, J., Lu, P., Zhu, Y., Xue, G., Li, M.: Toward secure multi keyword top-k retrieval over encrypted cloud data. IEEE Trans. Dependable Secure Comput. **10**(4), 239–250 (2013)
18. Sun, W., Wang, B., Cao, N., Li, M., Lou, W., Hou, Y.T., Li, H.: Privacy-preserving multi-keyword text search in the cloud supporting similarity-based ranking. In: Proceedings of ACM Symposium on Information, Computer and Communications Security, pp. 71–82 (2013)

19. Baek, J., Naini, R.S., Susilo, W.: Public key encryption with keyword search revisited. Proceedings of Computational Science and Its Applications (ICCSA). Lecture Notes in Computer Science, vol. 5072, pp. 1249–1259. Springer, Berlin/Heidelberg (2008)

20. Zhao, Y., Chen, X., Ma, H., Tang, Q., Zhu, H.: A new trapdoor indistinguishable public key encryption with keyword search. J. Wirel. Mobile Netw. Ubiquitous Comput. Dependable Appl. 3(1/2), 72–81 (2012)

21. Karnan, M., Akila, M., Krishnaraj, N.: Biometric personal authentication using keystroke dynamics: a review. Elsevier J. Appl. Soft Comput. 11(2), 1515–1573 (2011)

22. Li, C.T., Hwang, M.S.: An efficient biometrics based remote user authentication scheme using smart cards. J. Netw. Comput. Appl. 33(1), 1–5 (2010)

23. Giot, R., Abed, M.El., Hemery, B., Rosenberger, C.: Unconstrained keystroke dynamics authentication with shared secret. Elsevier Comput. Secur. 30(6–7), 427–445 (2011)

24. Upmanyu, M., Namboodiri,A. M., Srinathan, K., Jawahar,C.V. :Blind authentication: a secure crypto-biometric verification protocol. IEEE Trans. Inf. Forensics Secur. 5(2), 255–218 (2010)

25. Yeh, H.L., Chen, T.H., Hu, K.J., Shih, W.K.: Robust elliptic curve cryptography-based three factor user authentication providing privacy of biometric data. IET Inf. Secur. 7(3), 247–252 (2013)

26. Fan, C.I., Lin, Y.H.: Provably secure remote truly three-factor authentication scheme with privacy protection on biometrics. IEEE Trans. Inf. Forensics Secur. 4(4), 933–945 (2009)

27. Huang, X., Xiang, Y., Chonka, A., Zhou, J., Deng, R.H.: A generic framework for three-factor authentication: preserving security and privacy in distributed systems. IEEE Trans. Parallel Distrib. Syst. 22(8), 1390–1397 (2011)

28. Sun, W., Wang,B., Cao, N., Li, M., Lou, W., Hou, Y.T., Li, H, : Protecting your right: attribute-based keyword search with fine grained owner-enforced search authorization in the cloud. In: Proceedings of IEEE INFOCOM, pp. 226–234 (2014)

29. HL7 Version 3 Standard: Role-based Access Control Healthcare Permission Catalog, Re-lease 2, V3, RBAC. Accessed online at: http://www.hl7.org/implement /standards/ product_brief.cfm? prod-uct_id = 72

30. Sun, W., Wang, B., Cao, N., Li, M., Lou, W., Hou, Y.T., Li, H.: Verifiable privacy-preserving multi-keyword text search in the cloud supporting similarity-based ranking. IEEE Trans. Parallel Distrib. Syst. (TPDS) 25(11), 3025–3035 (2014)

31. HIPAA Security Guidance, Department of Health and Human Services, USA (2006). Accessed online at: http://www.hhs.gov/ocr/privacy/hipaa/administrative/securityrule/remoteuse.pdf

32. HL7 Version 3 Standard: Privacy, Access and Security Services (PASS)—Access Control. Release 1, PASS. Accessed online at: http://www.hl7.org/ implement/standards/ product_brief.cfm? prod-uct_id = 73

33. Bhattasali, T., Saeed, K., Chaki, N., Chaki, R.: Bio-authentication for layered remote health monitor framework. J. Med. Inf. Technol. 23(2014), 131–140 (2014)

34. Mayrhofer, R., Schmidtke,H.R., Sigg, S.: Security and trust in context-aware applications. Pers. Ubiquitous Comput. 18(1), 115–111 (2014)

35. Bhattasali, T., Chaki, R. Chaki, N. : Secure and trusted cloud of thing. In: Proceedings of IEEE India Conference, pp. 1–6 (2013)

36. Mahmoud, A., Niu, N.: Source code indexing for automated tracing. In: Proceedings of International Workshop on Traceability in Emerging Forms of Software Engineering, pp. 3–9 (2011)

37. Hu, H., Xu, J., Ren, C., Choi, B. : Processing private queries over untrusted data cloud through privacy homomorphism. In: Proceedings of IEEE International Conference Data Engineering (ICDE), pp. 601–612 (2011)

38. Habib, K., Torjusen, A., Leister, W.: A novel authentication framework based on bio-metric and radio fingerprinting for the IoT in eHealth. In: Proceedings of International Conference on Smart Systems, Devices and Technologies (SMART), pp. 32–37 (2014)

Part II
Systems Biology

Inference of Gene Regulatory Networks with Neural-Cuckoo Hybrid

Sudip Mandal, Goutam Saha and Rajat K. Pal

Abstract Current progress in cellular biology and bioinformatics allow researchers to get a distinct picture of the complex biochemical processes those occur within a cell of the human body and remain as the cause for many diseases. Therefore, this technology opened up a new door to the researchers of computer science as well as to biologists to work together to investigate the causes of a disease. One of the greatest challenges of the post-genomic era is the investigation and inference of the regulatory interactions or dependencies between genes from the microarray data. Here, a new methodology has been devised for investigating the genetic interactions among genes from temporal gene expression data by combining the features of Neural Network and Cuckoo Search optimization. The developed technique has been applied on the real-world microarray dataset of Lung Adenocarcinoma for detection of genes which may be directly responsible for the cause of Lung Adenocarcinoma.

Keywords Gene regulatory network · Microarray data · Neural network · Cuckoo search optimization

S. Mandal (✉)
ECE Department, GIMT, Krishna Nagar, India
e-mail: sudip.mandal007@gmail.com

G. Saha
IT Department, NEHU, Shillong, India
e-mail: dr_goutamsaha@yahoo.com

R.K. Pal
CSE Department, University of Calcutta, Kolkata, India
e-mail: pal.rajatk@gmail.com

© Springer India 2016
R. Chaki et al. (eds.), *Advanced Computing and Systems for Security*,
Advances in Intelligent Systems and Computing 395,
DOI 10.1007/978-81-322-2650-5_6

1 Introduction

Microarray technology [1, 2] has turned into one of the vital tools that many biologists use to observe genome-wide expression levels in a given living organism. A microarray consist of an array of glass slides on which particular DNA molecules corresponding to a unique gene are positioned in a systematic way at precise locations that are referred as features or spots. A microarray can have thousands of features, and each feature may include a million copies of identical DNA molecules or genes. Microarray profile falls into two categories, temporal or time series and static or classification-based data. The analysis of such data can reveal the unknown interactions and regulations among the genes inside the cell. Moreover, depending on the different gene expression levels, the status of a cell may be changed from normal to cancer. A gene can activate or inhibit expression level of other genes by different complex biological mechanisms. The graphical representation of these regulations among genes is known as a Gene Regulatory Network (GRN). Hence, identifying these regulations or interactions from microarray data with proper biological significance is a significant challenge to researchers.

The main problem of microarray technology is the 'curse of dimensionality,' as the size of microarray data is increasing exponentially with time. This is the base of problems for conventional statistical and mathematical algorithms. Nowadays, several 'intelligent' techniques, based on soft computing tools, have been implemented successfully to the problems of gene expression data analysis. The Neural-Cuckoo hybrid approach, which is described here, has been successfully used to discover sets of regulatory genes for each of the target genes from static and nontemporal DNA microarray data of Lung Adenocarcinoma. Lung cancer is a consequence of the extreme and uncontrolled expansion of cells in the tissues of lungs and is responsible for 1.3 billion annual deaths throughout the world. The main objective of this work is not to develop a complete and complex genetic network involving all regulatory genes for lung cancer, but to recognize the dependencies that exist among the dominant and responsible genes. Thus, this will lead to the consequent analysis of small-scale gene regulatory network.

Different statistical and mathematical models have been already developed for solving the inference problem of genetic networks in biological systems [3, 4]. Boolean networks [5, 6] look at binary state transition matrices to explore patterns in gene expression. The node of the network is either on or off depending on whether a signal surpasses a predetermined threshold level. On the other hand, Probabilistic Boolean Networks combine several promising Boolean functions together so that each one makes a contribution to the prediction of a target gene. A linear weighting network [7] has the benefit of simplicity since they use simple weight matrices to additively recombine the contributions of the different regulatory elements. A Bayesian network [8, 9] makes probabilistic transitions between network states considering conditional independence with no cycles in the network. However, cycles or loops are one of the most important mechanisms to ensure stability. A Dynamic Bayesian Network [10–12] merges the features of Hidden

Markov Models to include feedback. When modelling GRNs with the S-System [13, 14] method, the expression rates are defined by the difference between the two products of power–law functions, where the first denotes the activation term and the second denotes the degradation term of a gene product. Fuzzy work [15, 16] models the interactions between genes in gene regulatory pathways using fuzzy weights and clustering. In neural network [17, 18, 35] based model, genes are represented by nodes in the input and output layers of the neural network where the weight matrix between nodes represent the regulatory relationships between genes. Recurrent Neural Network [18, 19] is also a popular technique which is a closed loop neural network with a variable delay suitable to model system dynamics from temporal data. Moreover, hybrid methods such as dynamic evolutionary hybrid [20] Neural Network–Genetic Algorithm hybrid [21], Recurrent Neural Network-Particle Swarm hybrid [22, 23], Recurrent Neural Network-Fuzzy, Bee Colony Hybrid [24], S-system–Firefly hybrid [25], etc. have been already developed to reconstruct genetic networks from microarray data. Overall, among all machine learning approaches, the hybrid evolutionary approach is the most admired machine learning method for inferring correct connectivity of gene regulatory networks from microarray gene expression data.

On the other hand, Artificial Neural Network (ANN) is an effective soft computing tool for learning the pattern from the raw input data similar to the working principle of neurons in nervous system. Moreover, Cuckoo Search (CS) is a newly proposed nature-inspired optimization method that is motivated by Cuckoo's breeding behaviour to optimize a complex problem. Here, the characteristics of ANN and Cuckoo search optimization are hybridized to select the best combination of genes which are responsible for modifying the expression of each individual gene by minimizing the error between calculated and experimental gene expression value of output layer genes of the ANN. The detailed methodology is elaborated in Sect. 2, followed by experimental results. In the next section, conclusions regarding this hybridization process are discussed and references follow this section.

2 Methodology

The proposed Neural-Cuckoo hybrid approach utilizes a single-layer feedforward neural network to determine the weights of the edges between two consecutive layers of ANN with the help of perceptron-based learning method that minimizes the error between calculated and experimental output gene expression value. This ANN structure is hybridized with a new metaheuristic CS algorithm for creating hypothetical interconnections or regulatory edges among genes. Moreover, the maximum connectivity of each gene is restricted to N as it is observed that real-life GRN is sparsely connected, i.e., few genes participate in regulations. Before elaborating the proposed method, let us revisit some of the basic concepts regarding ANN and CS.

2.1 Preliminary of ANN and CS

Neural networks, specifically, are parallel processors, similar to the neurons in the human body, which have the capability of being trained through a supervised training or learning process. The common approach for networks is to accumulate them into different layers. Nodes within layers are connected by weighing factors called weights which denote the strength of the corresponding inputs. Layers of ANN can be classified as input, hidden, and output layers, respectively. The input layer is called as layer zero that stands for input attributes of the system. The output layer is used as the holders of output variables. The hidden layer is the intermediate layer between input and output layer to deal with nonlinearity of the training data. However, there is no limit to the number of hidden layers or number of nodes of those layers, although if the total number of nodes raises (and therefore number of weighted edges), computational efficiency decreases. Nodes of hidden layers accumulate the information flowing from previous layer and process these accumulated weighted upstream with the help of activation and threshold function. This processed information is sent to the next layer for further processing. During learning, the goal is to optimize the value of weighted connections of ANN structure such that for a particular set of input value, the corresponding pattern will appear as the output layer providing minimum amount of errors between learned and actual output. These networks are usually trained by different popular algorithms like perceptron rule, delta rules, back-propagation method, etc.

On the other hand, X. Yang first proposed Cuckoo Search optimization [26, 27] which is based on brood parasitism of cuckoo bird that reproduce their eggs by utilizing nests of others host birds. These birds have the ability to use other birds for raising their new generation. Cuckoo lay their eggs, one or more than one, in the nest of host birds in the absence of them. When the host bird recognizes the alien eggs, the host bird may destroy the eggs or can leave its nest and build a new nest with a certain probability P_a. To avoid this, cuckoos learn to make eggs similar to host bird's egg. However, the highest quality nest with eggs (i.e., best solutions) will be selected and move over to the next generation. This learning process may be considered as an optimization method. Levy flight [28] is an important characteristic of the Cuckoo search. Levy flight is defined as a random movement done by the birds with a step value of distributed probability. Next section describes the detailed process of the Neural-Cuckoo Hybrid.

2.2 Neural-Cuckoo Hybrid Approach

The Neural-Cuckoo search hybrid method which is proposed here combines a Cuckoo Search (CS) optimization with a supervised single-layer Artificial Neural Network (ANN) to create a 'hybrid' expert system that can find the genetic network. Here, one egg (solution) in a nest of cuckoo denotes a small set of genes that are chosen from the original set of genes in the microarray data. This set of genes comprise of the responsible regulatory genes that affect the target gene's expression value. The ANN is implemented to determine how fine the gene expression levels of those genes at a particular time point affect the target gene's expression value at next time instance. Here, the regulatory gene set are the input layer of ANN; target gene is the output node of ANN, and microarray dataset corresponding to these regulatory and target genes are used as training data for learning. For simplicity, each cuckoo place single egg at one of the randomly chosen nests whose number is predetermined.

The CS initializes with a population of different solutions or eggs on different random nest and the quality of each egg in host nest is calculated using ANN to observe its impact of regulatory genes on others gene in the database as part of one generation. By selecting proper CS parameters, the best quality solution can be obtained after iteration. Better host nest with better quality eggs (which is evaluated by the ANN) of each generation, will move in the next generations. After successful completion of all iterations, we have a set of regulatory genes which can affect the target gene most. The quality of host nest or fitness of a solution is simply proportional to the objective function that is the resultant error of ANN for the set of particular genes or nest. The overall GRN problem for all m genes is divided into m subproblems for the individual gene to determine optimal combinations of genetic interconnections of the network.

The proposed hybrid methodology is elaborated in detail below.

1. Let us consider that gene 1 (gene$_j$, where $j = 1$) is the first 'output' gene.
2. Apply the CS optimization to produce different combination of N input genes (gene$_2$; gene$_3$; ... ; gene$_N$) that will influence the 'output' gene (gene$_j$). It is considered that these selected set of N genes are only the responsible genes to modify the expression value of output gene. Auto regulation is also avoided. Each combination or set of genes is a host nest for the cuckoo, and all combinations of genes create the initial population of host nests. Each host nest of Cuckoo is equivalent to the solution of the problem for inference of GRN.
3. For each nest, choose expression values from the microarray database for gene$_2$; gene$_3$; ... ; gene$_N$ and gene$_j$ to generate a training dataset for ANN learning. Each pair in the training dataset contain the gene expression values for gene$_2$; ... ; gene$_N$ from the normal stage of the microarray data to be the input values to the Neural Network model, and the expression value for gene$_j$ from the cancer stage of the microarray data to be the target output of the Neural Network.
4. Use this training dataset to learn the ANN model with the help of the perceptron-based learning algorithm. It helps to find the optimum weights between the input layer (set of regulatory genes) and the output layer (target gene) until some stopping criterion (maximum iteration or minimum error of ANN) is achieved.
5. The final output error between calculated and experimental expression of ANN is returned as the fitness value (objective function) for that particular set of gene$_2$; gene$_3$; ... ; gene$_N$, i.e., host nest. Steps 3 and 4 are repeated for all nests in

a similar way. CS chooses the best nest with minimum objective function and returns the best solution.

6. Repeat Steps 2–5 as an ordinary CS iteration, using a standard parameter of Cuckoo search optimization on all nests to change the set of chosen input genes without varying the output gene.

7. CS ends after the stopping criterion (minimum value of error, i.e., objective function or maximum iteration) is achieved. The best nest with the highest quality of the egg (solution with the best set of genes that creates minimum error during ANN learning process) is selected. Hence, a set of regulatory genes corresponding to the target gene is obtained that affect the target gene most, and each regulation linking with the regulatory genes leads to an incoming interaction in the network structure.

8. For all genes, repeat the Steps 1–7 one after another by updating the value of j (gene$_j$).

After the completion of the NN-CS hybrid process, a set of regulatory genes for each gene is obtained. The corresponding genetic network structure can be inferred which contains a set of optimum incoming regulatory connections for each gene. A restriction is imposed to simplify the GRN structure that maximum of N incoming regulations is allowed for each gene of the network. This above stepwise process of the proposed method can also be visualized in Fig. 1.

3 Experimental Result

In this paper, the proposed method is tested on the microarray dataset of the Lung Adenocarcinoma (GEO Accession No.: GSE10072, which was obtained from NCBI [29] website) which has the data of 22,284 genes of 107 different persons. Therefore, finding the regulatory network between these huge numbers of genes is almost impossible, and the regulations are hard to interpret. Thus, in spite of developing a huge complex regulatory network, we should try to reconstruct dominant biological significant GRN for a small number of most responsible genes for Adenocarcinoma, and that will denote the major regulations during cancer. Now, Rough Set Theory can be used for dimensionality reduction of a dataset without losing important information. In earlier works [30, 31], Rule Reduction process using Rough Set was successfully implemented to identify only 15 most responsible genes for Lung Cancer from the huge database. Hence, prior knowledge of the expression value of the responsible genes is already available. The reduced dataset corresponding to the average value of the expression value for each of the genes for each stage is shown in Table 1. This dataset is used for training of ANN, where the average value of normal lungs and cancerous lungs is used as input and output layer of ANN, respectively.

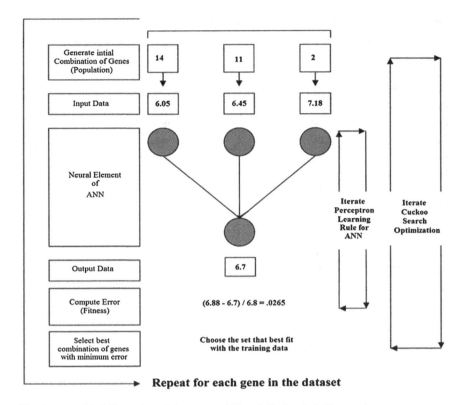

Fig. 1 A graphical illustration of the proposed Neural-Cuckoo hybrid execution

We have implemented the proposed methodology to the following reduced dataset of Table 1, which is used as training dataset using Matlab-7.6 on Windows 7 platform.

It is quite fascinating to find that average expression values of the maximum genes are decreasing due to cancer. The parameters for the CS and ANN are given in Tables 2 and 3, respectively. The resultant responsible gene set for each gene, i.e., Gene Regulatory Network is shown in Table 4.

Now, we can easily construct a directed graph according to the above table where each node or gene is connected to three other regulatory genes (shown in Table 4) with incoming edges. By accumulating, all nodes and their interaction GRN for Lung Adenocarcinoma is obtained which is shown in Fig. 2.

From the GRN, shown in Fig. 2, different direct or indirect influences can be inferred without difficulty. The parent genes have a direct causal influence on others genes. The intermediate genes particularly change the expression level of target genes via indirect influences. These can be considered as transcription factors of the gene network. The directed edges show the direction of regulations.

Table 1 The dataset used for learning neural network

Sl. no.	Gene ID	Average value of normal lungs	Average value of cancerous lungs
Gene1	201591_s_at	10.00	9.59
Gene2	201772_at	7.18	8.59
Gene3	201938_at	10.00	10.3
Gene4	202295_s_a	11.45	10.43
Gene5	203065_s_at	11.40	8.98
Gene6	203091_at	7.98	8.59
Gene7	203249_at	8.93	8.26
Gene8	205261_at	10.60	8.30
Gene9	206068_s_at	7.20	5.38
Gene10	208056_s_at	6.88	6.00
Gene11	209072_a	6.45	6.24
Gene12	209613_s_at	10.00	6.12
Gene13	218918_at	8.08	7.38
Gene14	222313_at	6.05	6.24
Gene15	49452_a	7.10	6.07

Table 2 CS parameter settings

Parameter	Value
Number of host nest/population	100
Discovery rate of alien eggs	0.25
Iteration	1500
Dimension/max interconnection	3

Table 3 ANN parameter settings

Parameter	Value
Learning parameter (α)	0.001
Minimum error	0.001
Maximum iteration	100

The above results are verified using the publicly available websites, namely 'The Database for Annotation, Visualization and Integrated Discovery (DAVID)' [32] and 'Gene Cards' [33]. These offer a complete set of functional annotation tools for scientist researchers to understand the biological significance of identified and studied genes.

Here, it has been observed from the gene ontological websites that among the 15 genes investigated, six genes are found to be directly related to Lung Adenocarcinoma (under investigation) and the remaining 10 genes are responsible for other types of cancers or related genetic diseases. This result suggests that the proposed methodology can extract biologically relevant information.

Table 4 The output of NN-CS hybrid approach for Lung Adenocarcinoma	Sl. no.	Set of regulatory genes obtained from NN-CS hybrid		
	Gene1	14	11	6
	Gene2	4	1	15
	Gene3	5	3	13
	Gene4	4	8	1
	Gene5	7	5	4
	Gene6	4	12	2
	Gene7	5	13	8
	Gene8	9	8	4
	Gene9	14	11	2
	Gene10	14	11	2
	Gene11	2	14	11
	Gene12	11	14	2
	Gene13	6	5	12
	Gene14	11	14	2
	Gene15	11	14	2

The pseudo code of this NN-CS hybrid is given as follows, where inputs are the training data (Table 1), and the outputs are set of regulatory genes for an individual gene.

Initialize the population with n = 100 host nests (solution) with dimension 3three (different gene number vary from 1 to 15);

While (t < max iteration (1500))

Randomly select a cuckoo (ith) by L'evy flights;

Calculate fitness value, i.e., quality of nest F_i of that solution;

Initialize the weight and learning rate for single-layer ANN; select input–output training data corresponding to input genes (according to solution or nest of CS) and output gene;

While (t1 < 100(max iteration) or error < 0.001 (stop criterion))

Calculate error between target and calculated output gene expression value;

Update weight according to Perceptron Learning rule;

End while;

Return error value as fitness value;

Randomly select another nest (jth) among others n host nests;

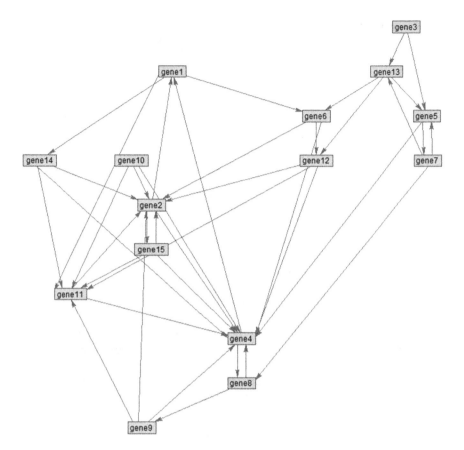

Fig. 2 Gene regulatory network by Neural-Cuckoo algorithm execution

```
if (Fj < Fi),
j is replaced by the new solution;
End;
Discard   worse   nests   with   a   fractional   probability
(pa = 0.25);
Generate new solution or new locations of host nest with the
help of Levy flights;
Keep the highest quality nest, i.e., best solution with
best fitness value;
Rank the available solutions and locate the current best;
end while;
Post-process and visualization of GRN;
```

4 Conclusion

A new Neural-Cuckoo hybrid approach for finding of genetic networks from microarray data has been illustrated throughout this paper. In addition to this, the NN-CS algorithm has been used to model genetic network by searching the best combination of regulatory genes that can affect a particular gene most. Moreover, as Cuckoo search is better [34] than the existing metaheuristic like Artificial Bee Colony (ABC), Differential Evolutionary (DE), and Particle Swarm Optimization (PSO), therefore Cuckoo Search Optimization is introduced for optimization problem of GRN structure. The method is applied to real-world microarray data of Lung Adenocarcinoma and it has been shown that it can infer the GRN successfully and efficiently so that it can precisely fit the training data by which it was already trained. So far as this study is concerned, this investigation is limited to classification of genes for a particular cause where input data is taken from microarray data available from different websites.

Here, for the verification of the results obtained, information from biological Ontological websites are used where investigation on the protein level functions of these genes are explored, and many correct inferences for cancer are recorded there. Though the biological significance of the interactions between genes for GRN of Lung Adenocarcinoma are needed to be validated in laboratory, we believe that this process has great potential in medical diagnosis, drug design, and soft computing perspective.

However, this process describes only the different regulations or dependencies among genes, but it does not specify the type (like activation or inhibition) and amount of inference or regulation. Moreover, during this process, the value of N is fixed and known, i.e., 3 but in practical cases, the number of regulations for a gene is unknown and not fixed. Therefore, there is always a chance of inclusion of false positive regulations in the network. Also, we neglect the self-regulation of genes for the simplicity. Therefore, in future, we need to incorporate all these points to improve the accuracy for the case of a real life network. Computational proficiency is also a significant characteristic of the proposed algorithm with respect to the number of genes in the microarray data. So, it is hoped that it can be easily implemented in large-scale gene regulatory networks with good efficiency on a standard computer.

One of the major problems in the inference of genetic network from microarray data is the shortage of known and state-of-the-art networks to validate and compare the proposed method. Therefore, this proposed method is needed to be applied to a known benchmark problem for identifying large complex networks from time series microarray to test its efficiency and accuracy. In the future, we shall try to modify the algorithm to incorporate all the parameters for successful modelling and validation of complex GRN.

References

1. National Center for Biotechnology Information (NCBI).: Microarrays: Chipping Away at the Mysteries of Science and Medicine, vol. 2004 NCBI, Bethesda (2004)
2. Masys, D.R.: Linking microarray data to the literature. Nat. Genet. **28**, 9–10 (2001)
3. Akutsu, T., Miyano, S., Kuhara, S.: Algorithms for inferring qualitative models of biological networks. In: Proceeding of Pacific Symposium on Biocomputing, 5, pp. 293–304 (2000)
4. Vijesh, N., Chakrabarty, S.K., Sreekumar, J.: Modelling of gene regulatory network: a review. J. Biomed. Sci. Eng. **6**, 223–231 (2013)
5. Liang, S., Fuhrman, S., Somogyi, R.: REVEAL, a general reverse engineering algorithm for inference of genetic network architectures. In: Proceeding of Pacific Symposium on Biocomputing, vol. 3, pp. 18–29 (1998)
6. Akutsu, T., Miyano, S., Kuhara, S.: Identification of genetic networks from a small number of gene expression patterns under the boolean network model. In: Proceeding of Pacific Symposium on Biocomputing, vol. 4, pp. 17–28 (1999)
7. Weaver, D.C., Workman, C.T., Stormo, G.D.: Modeling regulatory networks with weight matrices. In: Proceeding of Pacific Symposium on Biocomputing, vol. 4, pp. 112–123 (1999)
8. Drugan, M.M., Wiering, M.A.: Feature selection for Bayesian network classifiers using the MDL-FS score. Int. J. Approximate Reasoning **51**(6), 695–717 (2010)
9. Bielza, C., Larrañaga, P.: Discrete Bayesian network classifiers: a survey. ACM Comput. Surv. **47**(1, article 5), 1–43 (2014)
10. Murphy, K., Mian, S.: Modelling gene expression data using dynamic bayesian networks. In: Computer Science Division. University of California, Berkeley (1999)
11. Murphy, K.P.: Dynamic bayesian networks: representation, inference, and learning. In: Computer Science, p. 255. University of California, Berkeley (2002)
12. Perrin, B.E., Ralaivola, L., Mazurie, A., Bottani, S., Mallet, J., D'Alche-Buc, F.: Gene networks inference using dynamic Bayesian networks. Bioinformatics **19**(suppl. 2), II138–II148 (2003)
13. Wang, H., Quin, L., Dougherty, E.: Inference of gene regulatory network using S-system: a unified approach. In: Proceeding of 2007 IEEE Symposium CIBCB, pp. 82–89 (2007)
14. Nakayama, T., Seno, S., Takenaka, Y., Matsuda, H.: Inference of gene regulatory networks using immune algorithm. J. Bioinform. Comput. Biol. **9**, 75–86 (2011)
15. Du, P.P., Gong, J., Wurtele, E.S., Dickerson, J.A.: Modeling gene expression networks using fuzzy logic. IEEE Trans. Syst. Man Cybern. **35**, 1351–1359 (2005)
16. Dickerson, J.A., Cox, Z., Wurtele, E.S., Fulmer, A.W.: Creating metabolic and regulatory network models using fuzzy cognitive maps. In: Proceeding of North American Fuzzy Information Processing Conference (NAFIPS). Vancouver, B.C. (2001)
17. Vohradsky, J.: Neural network model of gene expression. FASEB J. **15**, 846–854 (2001)
18. Xu, R., Wunsch II, D., Frank, R.: Inference of genetic regulatory networks with recurrent neural network models using particle swarm optimization. IEEE/ACM Trans. Comput. Biol. Bioinform. **4**(4), 681–692 (2007)
19. Noman, N., Palafox, L., Iba, H.: Reconstruction of gene regulatory networks from gene expression data using decoupled recurrent neural network model. In: Natural Computing and Beyond (Springer), PICT6, pp. 93–103 (2013)
20. Ioannis, A.M., Andrei, D., Dimitris, T.: Gene regulatory networks modeling using a dynamic evolutionary hybrid. BMC Bioinform. **11**, 1–17 (2010)
21. Keedwell, E., Narayanan, A.: Discovering gene networks with a neural-genetic hybrid. IEEE/ACM Trans. Comput. Biol. Bioinform. **2**(3), 231–242 (2005)
22. Kentzoglanakis, K., Poole, M.: A swarm intelligence framework for reconstructing gene networks: searching for biologically plausible architecture. IEEE/ACM Trans. Comput. Biol. Bioinform. **9**(2), 355–371 (2012)

23. Xu, R., Wunsch II, D.C., Frank, R.L.: Inference of genetic regulatory networks with recurrent neural network models using particle swarm optimization. IEEE/ACM Trans. Comput. Biol. Bioinform. **4**(4), 681–692 (2007)
24. Rakshit, P., Das, P., Konar, A., Nasipuri, M., Janarthan R.: A recurrent fuzzy neural network model of a gene regulatory for knowledge extraction using invasive weed and artificial bee colony optimization algorithm. In: Proceeding of 1st International Conference on Recent Advances in Information Technology (RAIT) (2012)
25. Mandal, S., Saha, G., Pal, R.K.: S-system based gene regulatory network reconstruction using Firefly algorithm. In: Proceeding of Third International Conference on Computer, Communication, Control and Information Technology (C3IT), pp. 1–5 (2015)
26. Yang, X.S.: Nature-Inspired Metaheuristic algorithms, pp. 105–116, 2nd edn. Luniver Press, London (2010)
27. Jereesh, A.S., Govindan, V.K.: Gene regulatory network modeling using cuckoo search and S-system. Int. J. Adv. Res. Comput. Sci. Softw. Eng. **3**(9), 1231–1237 (2013)
28. Yang, X.S., Deb, S.: Cuckoo search via Lévy flights. In: Proceedings of World Congress on Nature & Biologically Inspired Computing (NaBIC 2009), pp. 210–214 (2009)
29. National Center for Biotechnology Information. http://www.ncbi.nlm.nih.gov
30. Mandal, S., Saha, G., Pal, R.K.: Reconstruction of dominant gene regulatory network from microarray data using rough set and bayesian approach. J. Comput. Sci. Syst. Biol. **6**(5), 262–270 (2013)
31. Mandal, S., Saha, G.: Rough set theory based automated disease diagnosis using lung adenocarcinoma as a test case. SIJ Trans. Comput. Sci. Eng. Appl. (CSEA) **1**(3), 59–66 (2013)
32. Database for Annotation, Visualization and Integrated Discovery. http://david.abcc.ncifcrf.gov
33. Gene Cards. http://www.genecards.org
34. Civicioglu, P., Besdok, E.: A conceptual comparison of the cuckoo-search, particle swarm optimization, differential evolution and artificial bee colony algorithms. Artif. Intell. Rev. 1–32 (2013)
35. Mandal, S., Saha, G., Pal, R.K.: Neural network based gene regulatory network reconstruction. In: Proceedings of Third International Conference on Computer, Communication, Control and Information Technology (C3IT), pp. 1–5 (2015)

Detection of Diabetic Retinopathy Using the Wavelet Transform and Feedforward Neural Network

Manas Saha, Mrinal Kanti Naskar and B.N. Chatterji

Abstract The early detection of diabetic retinopathy plays a significant role in modern ophthalmology. This experimental work presents the detection of diabetic retinopathy images by the implementation of the wavelet transform and feedforward neural network. The wavelet transform segments blood vessels from the retinal images and the changes in retinal vasculature due to diabetic retinopathy are characterized by the vessel features. The vessel features of all the input images are evaluated by the wavelet-based retinal image analyzer to tabulate the input and target databases. The input and target matrices are then fed to the neural network to detect the diabetic retinopathy images from the set of normal and diabetic retina images.

Keywords Diabetic retinopathy · Tortuosity · Multiresolution · Wavelet transform · Vasculature

M. Saha (✉)
Department of Electronics and Communication Engineering, Siliguri Institute of Technology, Siliguri, WB 734009, India
e-mail: manassaha77@yahoo.com

M.K. Naskar
Department of Electronics and Telecommunication Engineering, Jadavpur University, Kolkata, WB 700032, India
e-mail: mrinalnaskar@yahoo.co.in

B.N. Chatterji
Electronics and Communication Engineering Department, B. P. Poddar Institute of Management and Technology, Kolkata, WB 700052, India
e-mail: bnchatterji@gmail.com

© Springer India 2016
R. Chaki et al. (eds.), *Advanced Computing and Systems for Security*,
Advances in Intelligent Systems and Computing 395,
DOI 10.1007/978-81-322-2650-5_7

101

1 Introduction

One of the alarming health concerns in today's life is the diabetes. The different complications arising from diabetes are diabetic retinopathy affecting the vision, diabetic nephropathy troubling the kidney, and diabetic neuropathy distressing the nervous system. The diabetic retinopathy does not give any sharp symptom during its primitive stage. But often due to the high sugar level in the blood, it causes unheralded loss of vision. Therefore, the patient suffering from diabetes should undergo regular nonintrusive ophthalmic screening to check the occurrence and depth of this vision threatening disease. Fortunately, with the advancement of medical science it is found that the vasculature of the retinal blood vessels provide a clear insight to the presence and extent of several optical problems like diabetic retinopathy, glaucoma, hemorrhages, retinal artery occlusion, vein occlusion, neovascularization, and so on [1, 2].

The segmentation of blood vessels from the retinal images and subsequent investigation of their physical dimensions and arrangement pattern help to unearth a lot of information about diabetic retinopathy. As the number of blood vessels in the retina is very large and the image database of any clinic also contains too many images, the manual measurements become tiring, time consuming, and an erroneous process. Thus, the automated computer analysis is introduced for the segmentation and subsequent examination of the blood vessels from the retinal images.

In this paper, we extract the blood vessels of the input images consisting of the normal and diabetic retinopathy images by the multiresolution mathematical tool called the wavelet transform as shown in Fig. 1. The implementation of the wavelet transform using the Automated Retinal Image Analyzer (ARIA) [3] helps to segment the blood vessels from which different parameters like the number of vessel diameters (NVD), mean diameter (MD), standard deviation of diameter (SDD), minimum diameter (MND), maximum diameter (MXD), segment length (SL), diameter to length ratio (DLR), and tortuosity (TS) are obtained. Henceforth, these parameters (NVD, MD, SDD, MND, MXD, SL, DLR, TS) will be collectively considered as Parameters of Interest (POI). As will be discussed in Sect. 3, that during the mild nonproliferative stage of diabetic retinopathy, the blood vessels get swollen or balloon shaped resulting in the change of the physical dimensions of the blood vessels and vasculature pattern. Thus, the POI can suitably sense the diabetic retinopathy. The eight POIs of all the input images are computed by ARIA and two databases called the input and target databases are created. The input and the target databases are then fed to the classifier, the feedforward neural network. The input database is randomly divided into independent training, validation, and test datasets. The neural network is trained by the training data, so that it can accurately identify the diabetic retinopathy images from the mixed set of untrained test data. A sample input image from the DRIVE database [4] is shown in Fig. 2a and the wavelet segmented blood vessels of Fig. 2a is also shown in Fig. 2b for easy reference. The core contribution of the work is to exploit the correlation of the diabetic retinopathy with the corresponding retinal vascular morphology for the

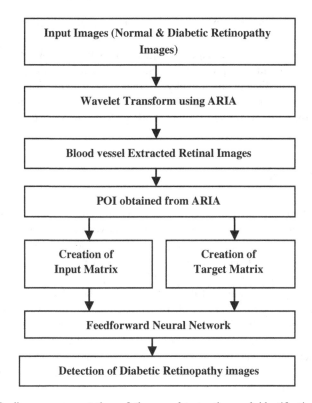

Fig. 1 Block diagram representation of the vessel extraction and identification of diabetic retinopathy images

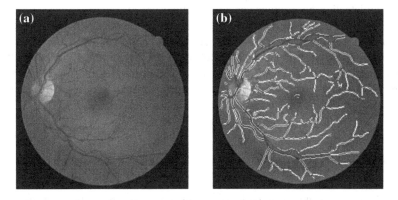

Fig. 2 **a** Sample retinal image and **b** blood vessel extracted retinal image of (**a**)

sake of retinal image discrimination using the wavelet transform and neural network.

The remaining part of the paper is organized as follows. A review of vessel segmentation techniques is presented in Sect. 2. Section 3 briefly discusses the clinical aspects of diabetic retinopathy. Retinal vessel extraction and its image discrimination are addressed in Sect. 4. Section 5 presents the experimental results. The conclusion with the scope of future work is drawn in Sect. 6.

2 Related Works

Over the last two decades, there has been a keen exploration on the segmentation of retinal blood vessels with numerous clinical objectives. Some of them as reported by Mendonca et al. [5] are macular avascular region detection, hypertensive retinopathy, diagnosis of cardiovascular diseases, etc. Chaudhuri et al. [6] proposed the detection of the retinal blood vessels with the help of the 2-D operators based on the spatial and optical features of the blood vessels. The ridge based retinal vessel segmentation proposed by Staal et al. [4] emphasizes on the extraction of image ridges coinciding closely with the centerlines of the blood vessels. Soares et al. [7] also reported the automated segmentation of retinal blood vessels based on the 2-D gabor wavelet and supervised classification. The graph cut technique used by Gonzalez et al. [1] segmented the retinal vascular tree from the retina image which was later used to trace the location of the optic disk. As the matched filter responds equally to vessels and non vessel edges, leading to incorrect vessel detection, Zhang et al. [2] suggested a novel matched filter approach for retinal vasculature extraction. A valuable algorithm combining the differential filters with the morphological operators for the automated detection of retinal vascular tree is presented in [5]. The curvelet transform was deployed by Esmaeili et al. [8] for the contrast enhancement and edge detection of the poor contrasted retinal vessels.

The pixel-based classification for detecting retinal blood vessels was addressed by Kharghanian et al. [9] where the classifiers, Bayesian and Support Vector Machine are used to segregate each pixel of the retina as "vessel" pixel and "non vessel" pixel. In [9], the response of both the gabor wavelet transform at different scales and the orthogonal line operators together with the gray level of the green channel constitute the image feature vector needed for the classification.

3 Diabetic Retinopathy

The diabetic retinopathy is an optical complication caused by diabetes. About 10 % of diabetic patients have the probability of developing diabetic retinopathy [3]. It is shocking to learn that about 347 million people are diabetic worldwide [10]. The four stages of this ocular disease are mild nonproliferative retinopathy, moderate nonproliferative retinopathy, severe nonproliferative retinopathy, and proliferative retinopathy [11]. Mild nonproliferative retinopathy is the first stage of diabetic retinopathy and the patient may not notice the change in his vision until and unless it is monitored by automated regular screening. The tiny retinal blood vessels swell like balloons and are clinically known as microaneurysyms. During moderate nonproliferative retinopathy, some blood vessels providing nourishment to the cells of the retina get blocked. The severe nonproliferative retinopathy is marked by the blockage of too many blood vessels. Therefore, a large portion of the retina gets deprived from the blood supply. The under nourished areas of the retina activate the growth of the new blood vessels. Proliferative retinopathy is the matured stage of retinopathy. The blood vessels not only grow along the surface of the retina but also through the vitreous gel to provide blood to the less nourished parts of the retina. The walls of the newly grown blood vessels are thin and fragile. So, blood leaks from the walls of the vessels into the center of the eye causing ultimate loss of vision.

4 Retinal Vessel Extraction and Its Image Classification

The complete work represented by the block diagram of Fig. 1 is carried out in Matlab environment with version: 7.14.0.739 (R2012a). The two major steps are (a) vessel extraction by the wavelet transform and (b) detection of diabetic retinopathy images by the feedforward neural network.

4.1 Vessel Extraction by the Wavelet Transform

Here, 20 normal and 12 diabetic retinopathy images are taken from the DRIVE [4] and REVIEW [12] databases, respectively. These input images are passed through the ARIA [3]. The ARIA is a software used for the automatic recognition and quantification of blood vessels present in the retina. The input images are subjected to histogram based masking for improving enhancement and accuracy. Then, the blood vessels of the retina are segmented with the help of wavelet transform. The wavelet transform used here is the Isotropic Undecimated Wavelet Transform (IUWT). The ARIA with IUWT is selected here, because the working retinal images are almost isotropic in nature. The wavelet coefficients, W_{j+1} generated by IUWT can be expressed from [3] as

$$W_{j+1} = C_j - C_{j+1} \tag{1}$$

where C_j and C_{j+1} are the side by side sets of the scaling coefficients. The group of wavelet coefficients generated at each iteration is referred to as the wavelet level. The blood vessels of the preprocessed retinal images are segmented by the addition of these wavelet levels which present the highest level of contrast and conditional thresholding [3]. The wavelet function used here is determined as difference between the two levels of resolution and can be expressed from [13] as

$$\frac{1}{4}\psi\left(\frac{x}{2},\frac{y}{2}\right) = \phi(x,y) - \frac{1}{4}\phi\left(\frac{x}{2},\frac{y}{2}\right) \tag{2}$$

where $\phi(x,y) = \phi_1(x)\phi_1(y)$, where

$$\phi_1(x) = \frac{1}{12}\left(|x-2|^3 - 4|x-1|^3 + 6|x|^3 - 4|x+1|^3 + |x+2|^3\right)$$

where $\phi_1(x)$ is the third order spline. The post wavelet filtering techniques like the implementation of the connectivity constraint for edge detection are followed for optimal image conditioning. Once the blood vessels are extracted, as shown in Fig. 2b, the eight POIs are obtained to act as the feature vector of the input matrix. The 8×32 input matrix represents the 32 (20 normal and 12 diabetic retinopathy) image samples of 8-dimension feature vector. The 2×32 target matrix is used to train the feedforward neural network with either diabetic retinopathy positive or negative cases. After the creation of the two databases, the matrices are fed to the neural network for classification.

4.2 Detection of the Diabetic Retinal Images by the Feedforward Neural Network

The feedforward neural network based on conjugate gradient back propagation algorithm is a classifier with three layers—the input layer, the hidden layer, and the output layer.

The steps of classifying include the selection of the input and target matrices, network creation with suitable number of neurons in the hidden layer, training the created network, and finally the performance evaluation of the classifier based on the mean square error (MSE) and confusion matrix.

Once the input and target matrices are presented to the network as discussed in Sect. 4.1, the input matrix is randomly divided by the feedforward neural network into three distinct and independent datasets, namely the training, validation, and testing datasets. The number of training, validation, and testing samples in each dataset is obtained by considering 70, 15, and 15 % of the input samples. The training samples are used to train the network while the validation and the testing samples help to measure the network generalization and performance respectively.

Finally, the classification of the retinal images into normal and diabetic retinopathy images is observed from the performance plot and the confusion matrices as shown in Figs. 3 and 4 respectively.

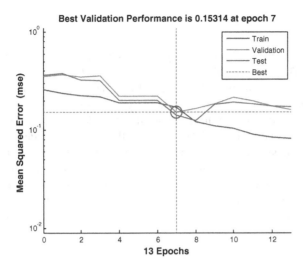

Fig. 3 Performance plot of the feedforward neural network

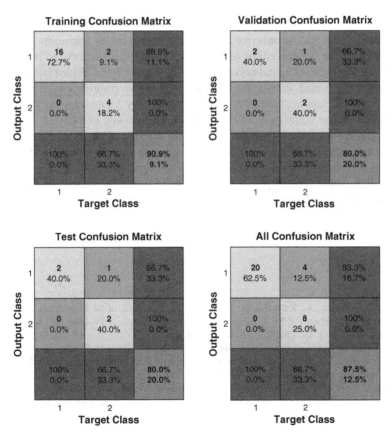

Fig. 4 Confusion matrices obtained from the feedforward neural network

5   Result and Discussion

The performance plot as shown in Fig. 3 gives the training, validation, and test performance. The best validation performance is 0.15314 at epoch 7. The different confusion matrices: training, validation, and test, and All Confusion matrices are shown together in Fig. 4. With 10 neurons present in the hidden layer, the All Confusion matrix (not shown here) shows that 84.4 % of retinal images are correctly classified as normal and diabetic retinopathy images. In order to increase the correct rate of classification, the number of neurons in the hidden layer is increased from 10 to 15. With 15 neurons in the hidden layer, the percentage of correct classification is increased from 84.4 to 87.5 % which is shown in Fig. 4. Similar percentage of correct training, validation, and testing can be understood from the training, validation, and testing confusion matrices as shown in Fig. 4.

6   Conclusion and Scope of Future Work

This investigation helps to detect this optical disease from the retinal images. Using the ARIA, the blood vessels are extracted by the wavelet transform. The POI constitutes of the feature vector for classification by the feedforward neural network. The feedforward neural network with 10 neurons in the hidden layer screens the input images and detects the diabetic retinopathy positive images with the correct rate of classification of 84.4 %. Later on, by increasing the number of neurons to 15, the correct rate of classification is improved to 87.5 %.

The present work deals with the first stage of diabetic retinopathy. However, the most critical stage called the proliferative retinopathy can be identified by incorporating the concept of power spectral density of the retinal images as a feature vector. This would definitely help the clinicians to detect the diabetic retinopathy at both the early and later stages. This work can also be extended to screen the same in children.

References

1. Salazar-Gonzalez, A., Kaba, D., Li, Y., Liu, X.: Segmentation of the blood vessels and optic disc in retinal images. IEEE J. Biomed. Health Inform. **18**(6), 1874–1886 (2014)
2. Zhang, B., Zhang, L., Zhang, L., Karray, F.: Retinal vessel extraction by matched filter with first-order derivative of Gaussian. Comput. Biol. Med. **40**, 438–445 (2010)
3. Bankhead, P., Scholfield, C.N., McGeown, J.G., Curtis, T.M.: Fast retinal vessel detection and measurement using wavelets and edge location refinement. PLoS ONE **7**(3), e32435 (2012)
4. Staal, J., Abràmoff, M.D., Niemeijer, M., Viergever, M.A., Ginneken, B.V.: Ridge-based vessel segmentation in color images of the retina. IEEE Trans. Med. Imaging **23**(4), 501–509 (2004)

5. Mendonca, A.M., Campilho, A.: Segmentation of retinal blood vessels by combining the detection of centerlines and morphological reconstruction. IEEE Trans. Med. Imaging **25**(9), 1200–1213 (2006)

6. Chaudhuri, S., Chatterjee, S., Katz, N., Nelson, M., Goldbaum, M.: Detection of blood vessels in retinal images using two-dimensional matched filters. IEEE Trans. Med. Imaging **8**(3), 263–269 (1989)

7. Soares, J.V.B., Leandro, J.J.G., Cesar, R.M., Jelinek, H.F., Jelinek, M., Cree, M.J.: Retinal vessel segmentation using the 2-d gabor wavelet and supervised classification. IEEE Trans. Med. Imaging **25**(9), 1214–1222 (2006)

8. Esmaeili, M., Rabbani, H., Mehri, A., Dehghani, A.: Extraction of retinal blood vessels by curvelet transform. In: Proceedings of 16th IEEE International Conference on Image Processing, ICIP 2009, pp. 3353–3356, Cairo, Egypt, 7–10 Nov 2009

9. Kharghanian, R., Ahmadyfard, A.: Retinal blood vessel segmentation using gabor wavelet and line operator. Int. J. Mach. Learn. Comput. **2**(5) (2012)

10. http://www.who.int/mediacentre/factsheets/fs312/en/

11. http://www.nei.nih.gov/health/diabetic/diabeticretino.pdf

12. Al-Diri, B., Hunter, A., Steel, D., Habib, M., Hudaib, T., Berry, S.: Review—a reference data set for retinal vessel profiles. In: The 30th Annual Conference of the IEEE Engineering in Medicine and Biology Society in Vancouver, British Columbia, Canada, 20–24 Aug 2008

13. Starck, J.L., Fadili, J., Murtagh, F.: The undecimated wavelet decomposition and its reconstruction. IEEE Trans. Image Process. **16**(2), 297–309 (2007)

Liver Fibrosis Diagnosis Support System Using Machine Learning Methods

Tomasz Orczyk and Piotr Porwik

Abstract Liver fibrosis is a common disease of the European population (but not only them). It may have many backgrounds and may develop with a different rapidity—it may stay hidden for many years or rapidly develop into terminal stage called cirrhosis, where liver can no longer fulfill its function. Unfortunately, current methods of diagnosis are either connected with a potential risk for a patient and require a hospitalization or are expensive and not very accurate. This paper presents a comparative study of various feature selection algorithms combined with selected machine learning algorithms which may be used to build an advanced liver fibrosis diagnosis support system based on a nonexpensive and safe routine blood tests. Experiments carried out on a dataset collected by authors, proved usability and satisfactory accuracy of the presented algorithms.

Keywords Machine learning · Ensemble classifier · Hepatology · Liver fibrosis

1 Introduction

Liver diseases are one of the leading causes of death in many world countries. Most common causes of chronic liver disease include: alcoholism, viral hepatitis, metabolic, fatty liver diseases, etc. [1, 2]. A more serious form of liver disease may progress to cirrhosis and finally lead to end-stage disease. This type of disease is regarded as a civilization threat. Recognition of the disease severity/stage is difficult even for experienced physicians. For this reason, reliable medical support systems,

T. Orczyk (✉) · P. Porwik
University of Silesia in Katowice, Institute of Computer Science,
Bedzinska 39, Sosnowiec, Poland
e-mail: tomasz.orczyk@us.edu.pl

P. Porwik
e-mail: piotr.porwik@us.edu.pl

© Springer India 2016
R. Chaki et al. (eds.), *Advanced Computing and Systems for Security*,
Advances in Intelligent Systems and Computing 395,
DOI 10.1007/978-81-322-2650-5_8

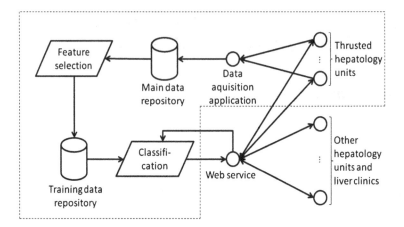

Fig. 1 General concept of proposed medical diagnosis support system (*dotted line* represents currently developed areas)

which can support medical centers, are increasingly introduced in many domains of medicine.

In liver diagnosis, clinicians use various medical data. One of them is the liver biopsy. Unfortunately, liver biopsy is an invasive procedure, which may cause severe health complications and its accuracy is limited by sampling heterogeneity and observer experience. More convenient for patients are noninvasive alternatives, including transient elastography and blood tests. These methods are relatively new but extensively validated [3–6].

Evaluation of the liver fibrosis stage is crucial for the therapeutic decisions. Therefore, we propose a basis for the new medical support system which makes a decision of fibrosis stage on the basis of the patient's blood parameters only.

The early stage of classifier-based medical diagnosis support system is currently tested in daily clinical practice in hospital's hepatology department with the cooperation of physicians (see Fig. 1).

2 Data Characteristics

Data used in this experiment comes from the Gastroenterology and Hepatology Department of the Independent Public Central Hospital of the Silesian Medical University. It contains medical records of 290 patients infected with a hepatitis virus type C. These records consist of patient's age, routine blood test results, and the liver biopsy result (see Table 2).

In the original database, the liver fibrosis stage was labeled according to the five step (F0...F4) *METAVIR* scale [7], but due to overlapping of neighboring classes and some level of uncertainty of biopsy results (subjective assessment of

Table 1 Classes (re) assignment

Class	METAVIR score	Count (%)
L	F0, F1	129 (44.5)
M	F2, F3	102 (35.2)
H	F4	59 (20.3)
Total		290 (100.0)

histopathologist) [8, 9], after a medical consultation, the number of classes used in the experiments has been narrowed to three (see Table 1). New classes represent low, medium, and high advancement of the liver fibrosis.

3 Machine Learning

3.1 Feature Selection

Due to the high dimensionality of the input data, a preprocessing stage has been introduced. In the first step, a feature selection is performed. Selection of data can be carried out by different techniques, which will be explained later. Selected features are treated as input data for various classification algorithms. It is a well know fact that limiting the feature space before the classification stage may help to gain the classification accuracy [10, 11], reduces the computational cost, and also minimizes the effect of the curse of dimensionality. Seven different feature selection algorithms (briefly described below) have been tested and compared with classification made on a complete set of features (labeled as "All") and on a subset of features proposed by medical experts. Below is a list of tested feature selection algorithms, together with a brief description of them

CFS evaluates the worth of a subset of attributes by considering the individual predictive ability of each feature along with the degree of redundancy between them. Subsets of features that are highly correlated with the class while having low intercorrelation are preferred by this algorithm [12].

ReliefF evaluates the worth of an attribute by repeatedly sampling an instance and considering the value of the given attribute for the nearest instance of the same and different class. It can operate on both discrete and continuous class data [13–15]. Algorithm has been set to choose 10 attributes.

SVM evaluates the worth of an attribute by using an *SVM* classifier. Attributes are ranked by the square of the weight assigned by the *SVM*. Attribute selection for multiclass problems is handled by ranking attributes for each class separately using a one-vs-all method and then taking from the top of each "pile" to give a final ranking [16].

Genetic Wrapper (IBk and J48) evaluates attribute sets by using a learning scheme. Cross-validation is used to estimate the accuracy of the learning scheme for a set of attributes [17]. This method have been used together with the *IBk* (*k-NN*)

and the *J48* (*C4.5*) classification algorithms. For optimal subset selection, a genetic search algorithm has been used [18].

SingleSeparate is an author proposed method that evaluates a worth of an attribute by using tenfold cross-validation using *IBk* classifier operating on a single attribute based on a *one-vs-all* validation method separately for each class and finally selects four attributes that best distinguishes each of the classes.

SingleAccuracy evaluates a worth of an attribute by using tenfold cross-validation using *IBk* classifier and evaluates classification overall accuracy based on a single attribute. This method selects 10 attributes which individually gives highest classification accuracy [19].

3.2 Classification

The object classification methods are valuable data analysis tools for multiparameter data sets and complex problems. These methods may be used to build a medical diagnosis support system [20].

A classifier is an algorithm that takes a set of parameters (features) that characterize objects and uses them to determine the type (class) of each object. The object together with its features can be represented as a data vector, $f \in F$. The classification algorithm Ω maps the feature space F of the class label C according to the mapping

$$\Omega_j : f^i \to C, \tag{1}$$

where $i = 1,\ldots,9$ represents various feature selection methods and $j = 1,\ldots,5$ means number of classification algorithms used in experiments.

The mapping (1) is performed on the basis of a set of the learning examples given by a domain expert. Our expert is a medical pathologist. Mentioned examples consist of observed features which describe the object and its correct classification given by the expert.

After the learning stage, a classifier is ready for recognition and classification of objects. Objects, in our case, are the patients with different liver fibrosis stages (see Table 2). Our task is to indentify patient's liver fibrosis stage based on his/her blood test analysis results.

For the purpose of this study, five most popular classification algorithms have been tested for using them in the medical diagnosis support system

J48 Pruned *C4.5* decision tree [21]. *C4.5* is based on the "Top Down Induction of Decision Tree." The *C4.5* algorithm belongs to the *ID3* family which uses the information gain that measures how well the given attribute separates the training examples according to the target classification.

IBk *k*-nearest neighbors classifier [22]. The *k-Nearest Neighbors* classifier is assuming membership of a new object to a class on the basis of comparing it against

Table 2 Patients' data characteristics

No.	Parameter (unit)	Mean	Std. deviation
1	Age (years)	57.4	14.15
2	Hemoglobin (g/l)	14.6	1.71
3	RBC (10^6/µl)	4.8	0.62
4	WBC (10^3/µl)	6.1	1.90
5	PLT (10^3/µl)	197.1	59.50
6	PT (s)	12.0	4.70
7	PTP (%)	99.6	15.75
8	APTT (s)	33.5	5.59
9	INR	1.0	0.11
10	ASPT (IU/l)	63.8	48.54
11	ALAT (IU/l)	82.5	64.26
12	ALP (IU/l)	80.3	29.99
13	Bilirubin (mg/dl)	1.0	0.64
14	GGTP (IU/l)	70.9	66.15
15	Creatinine (mg/dl)	1.0	0.35
16	Glucose (mg/dl)	96.4	19.83
17	Na (mmol/l)	138.3	3.10
18	K (mmol/l)	4.3	0.46
19	Cholesterol (mg/dl)	187.0	38.71
20	Total Protein (g/dl)	7.5	0.64
21	Albumins (g/dl)	0.5	0.25
22	Albumins (%)	60.9	5.92
23	α1 Globulins (%)	2.7	0.87
24	α2 Globulins (%)	9.2	1.53
25	β Globulins (%)	10.6	1.70
26	γ Globulins (%)	16.4	5.09

RBC red blood cells; *WBC* white blood cells; *PTL* platelets; *PT* prothrombin time; *PTP* prothrombin ratio; *APTT* activated partial thromboplastin time; *INR* international normalized ratio; *AST* aspartate aminotransferase; *ALT* alanine aminotransferase; *ALP* alkaline phosphatse; *GGT* 7-glutamyltransferase; *Na* natrium; *K* kalium

a set of sample (prototype) objects. During classification, a voting process of *k*-nearest neighbors is used.

RandomForest forest (ensemble) of random tree classifiers [23]. Random Forests are built up from the ensemble of *Random Trees* which, in contrary to classic decision trees, are built using randomly selected subsets of features for each node.

OneR uses the minimum error attribute for prediction, discretizing numeric attributes [24]. *OneR*, short for "One Rule," is a classification algorithm that generates one rule for each attribute in the data and then selects the rule with the smallest total error as its "one rule." It uses a frequency table for each attribute against the target to create a rule for an attribute.

DecisionTable simple decision table majority classifier [25]. It is a decision table with a default rule mapping to the majority class.

Each algorithm has been tested in its basic implementation as well as in ensemble of single parameter classifiers proposed by authors. Division of the original feature space into a single dimensional subspace has been made mainly due to the fact that analyzed dataset contains a significant amount (~ 25 %) of randomly missing attribute values within each patient's record.

Ensemble architecture has been described in detail in [19]. It is an ensemble of classifiers of the same type, working on a single feature. The decision of the ensemble is achieved by summarizing support values for each class, returned by each classifier in the ensemble. If there is a missing attribute in the input vector, classifiers based on this attribute returns support value of 0 for all classes. In the learning process, each of the classifiers from the ensemble is trained only on a real data, if an attribute is missing, the training record is not formed for this classifier and thus no artificially generated data needs to be induced. Finally, due to the fact that each feature is analyzed separately, there is no need to normalize data prior to training or classification.

In order to obtain most accurate results and use all available records of data, classification accuracy has been measured in a leave-one-out cross-validation process on the training set. The experiments have been done in the KNIME [26] environment with the use of WEKA [27] components.

4 Results

Features selected by the feature selection algorithms for use in the classification stage have been presented in Table 3.

Complete results of the experiment have been shown in Fig. 2. The three best classifiers which are close to the accuracy of 70 % are: J48 using CFS feature selection, Random Forest using ReliefF feature selection, and ensemble of single parameter IBk classifiers using SingleSeparate feature selection method. It is worth to notice that best results were obtained on the smallest feature sets (with the lowest number of attributes).

Results may look comparative, but what's significant for them is the fact that the simple *IBk* has not obtained best overall accuracy for any feature selection method, while ensemble of single feature-based *IBk* classifiers have improved overall accuracy for all tested feature selection methods. So, it may be stated, that this modification significantly improved the *IBk* classifier accuracy, what can be clearly seen on Fig. 3.

Table 3 Attributes chosen by feature selection algorithms

No.	Parameter	CFS	ReliefF	SVM	Gen IBk	Gen C45	SS IBk	SA IBk	Phys
1	Age	X	X	X	X	X	X	X	
2	Hemoglobin					X			
3	RBC			X	X				
4	WBC			X	X				X
5	PLT	X	X	X	X	X	X	X	X
6	PT								X
7	PTP	X	X		X	X	X		X
8	APTT		X		X	X			X
9	INR					X	X	X	X
10	ASPT	X	X		X	X	X	X	X
11	ALAT		X	X	X	X	X	X	X
12	ALP		X	X	X				X
13	Bilirubin			X	X				X
14	GGTP	X	X	X		X	X	X	
15	Creatinine			X					
16	Glucose								
17	Na	X	X	X	X	X			
18	K				X	X			
19	Cholesterol					X	X	X	
20	Total Protein				X				
21	Albumins								
22	Albumins %	X				X	X		X
23	α1 Globulins							X	X
24	α2 Globulins		X						X
25	β Globulins				X				X
26	γ Globulins					X		X	X
Total		7	10	10	14	14	9	9	15

All of the tested classifiers tend to underestimate liver fibrosis stage, the most balanced results were obtained for *Random Forest* classifier. Also, different classifiers deal differently with different classes, the ensemble *IBk* best recognizes the class **L**, the class **M** is best recognized ex aequo by ensemble *IBk* and *Random Forest*, while the class **H** is best recognized by the Random Forest. The *J48* classifier which achieved the best overall accuracy was not best at recognizing any class.

Table 4 presents confusion tables for the three best classifiers (a: *J48 + CFS*, b: *Random Forest + ReliefF*, c: *ensemble IBk + SingleSeparate*).

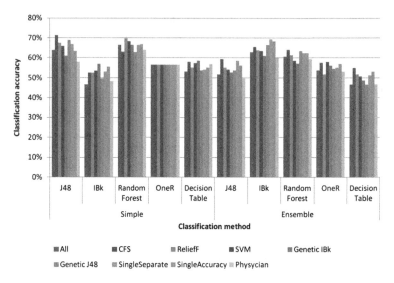

Fig. 2 Feature selection methods comparison

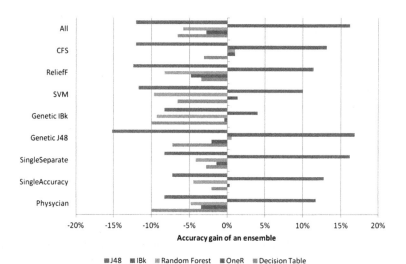

Fig. 3 Simple versus ensemble classifier

Table 4 Confusion tables for best classifiers

(a)		Classified		
		L	M	H
Actual	L	111	13	5
	M	35	58	9
	H	7	14	38
(b)		Classified		
		L	M	H
Actual	L	97	25	7
	M	29	63	10
	H	7	10	42
(c)		Classified		
		L	M	H
Actual	L	113	16	0
	M	37	63	2
	H	20	14	25

5 Conclusions

Dividing a feature space into subspaces containing single attributes resulted in a gain of classification accuracy, but only for the simplest k-nearest neighbors classifier. What is significant, ensemble based on this classifier became one of the best classifiers in the comparison, while basic version of this classifier was the weakest one. Additionally, as expected, feature reduction have gained the classification accuracy in a vast majority of analyzed cases, but what is also important, there was no best feature selection algorithm for all classifiers. From the confusion tables of best tested classifiers, it can be seen that different classifiers, operating on different subsets have a different best and worst distinguishable classes. Because of this fact, it may be possible to further correct the classification accuracy by creating an ensemble of different classifiers, and a combination rule that would use weights to promote single classifiers decision in the respect of the recognized class and best recognized class by this type of a classifier. Especially for k-NN based classifiers, the unbalanced classes in the training set may be a problem, which needs to be further examined. Also, the further tuning of the ensemble *IBk* is possible by the introduction of additional weighting of classifiers' decisions. Overall accuracy results of proposed ensemble classifiers shows that it is possible to build a reliable decision module of a medical diagnosis support system using presented feature selection and classification algorithms.

Acknowledgments This work was supported by the Polish National Science Center under the grant no. DEC-2013/09/B/ST6/02264.

References

1. Wojtyniak, B., Goryński, P., Moskalewicz, B.: Sytuacja zdrowotna ludności polski i jej uwarunkowania. Technical report, Narodowy Instytut Zdrowia Publicznego-Państwowy Zakład Higieny, 2012 (in Polish)
2. Stevenson, M., Lloyd-Jones, M., Morgan, M.Y., Wong, R.: Non-invasive diagnostic assessment tools for the detection of liver fibrosis in patients with suspected alcohol-related liver disease: a systematic review and economic evaluation. Health Technol. Assess. **16**(4) (2012). doi:10.3310/hta16040
3. Lucas, P.J.F., Segaar, R.W., Janssens, A.R.: HEPAR: an expert system for diagnosis of disorders of the liver and biliary tract. Liver **9**, 266–275 (1989)
4. Adlassnig, K.P., Horak, W.: Development and retrospective evaluation of HEPAXPERT—I: a routinely-used expert system for interpretive analysis of hepatitis A and B serologic findings. Artif. Intell. Med. **7**, 1–24 (1995)
5. Zhao, Y.K., Tsutsui, T., Endo, A., Minato, K., Takahashi, T.: Design and development of an expert system to assist diagnosis and treatment of chronic hepatitis using traditional Chinese medicine. Med. Inform. **9**, 37–45 (1994)
6. Shiomi, S., Kuroki, T., Jomura, H., Ueda, T., Ikeoka, N., Kobayashi, K., Ikeda, H., Ochi, H.: Diagnosis of chronic liver disease from liver scintiscans by fuzzy reasoning. J. Nucl. Med. **36**, 593–598 (1995)
7. Bedossa, P., Poynard, T.: An algorithm for the grading of activity in chronic hepatitis C. The METAVIR Cooperative Study Group. Hepatology **24**, 289–293 (1996)
8. Regev, A., Berho, M., Jeffers, L., Milikowski, C., Molina, E., Pyrosopoulos, N., Feng, Z., Reddy, Z., Schiff, E.: Sampling error and intraobserver variation in liver biopsy in patients with chronic HCV infection. Am. J. Gastroenterol. **97**(10), 2614–2618 (2002)
9. Bedossa, P., Dargere, D., Paradis, V.: Sampling variability of liver fibrosis in chronic hepatitis C. Hepatology **38**, 1449–1457 (2003)
10. Doroz, R., Porwik, P.: Handwritten signature recognition with adaptive selection of behavioral features. In: Communications in Computer and Information Science (CISIM), vol. 245, pp. 128–136. Springer, Kolkata (2011)
11. Porwik, P., Doroz, R.: Self-adaptive biometric classifier working on the reduced dataset. In: Hybrid Artificial Intelligence Systems. Lecture Notes in Computer Science, vol. 8480, pp. 377–388. Springer International Publishing, New York (2014)
12. Hall, M.A.: Correlation-Based Feature Subset Selection for Machine Learning. Hamilton, New Zealand (1998)
13. Kira, K., Rendell, L.A.: A practical approach to feature selection. In: Ninth International Workshop on Machine Learning, pp. 249–256 (1992)
14. Kononenko, I.: Estimating attributes: analysis and extensions of RELIEF. In: European Conference on Machine Learning, pp. 171–182 (1994)
15. Robnik-Sikonja, M., Kononenko, I.: An adaptation of Relief for attribute estimation in regression. In: Fourteenth International Conference on Machine Learning, pp. 296–304 (1997)
16. Guyon, I., Weston, J., Barnhill, S., Vapnik, V.: Gene selection for cancer classification using support vector machines. Mach. Learn. **46**, 389–422 (2002)
17. Kohavi, R., John, G.H.: Wrappers for feature subset selection. Artif. Intell. **97**(1–2), 273–324 (1997)
18. Goldberg, D.E.: Genetic Algorithms in Search, Optimization and Machine Learning. Addison-Wesley, Reading (1989)
19. Orczyk, T., Porwik, P., Bernaś, M.: Medical diagnosis support system based on the ensemble of single-parameter classifiers. J. Med. Inform. Technol. **23**(2014), 173–179 (2014)
20. Foster, K.R., Koprowski, R., Skufca, J.D.: Machine learning, medical diagnosis, and biomedical engineering research—commentary. Biomed. Eng. **13**(94) (2014). doi:10.1186/1475-925X-13-94

21. Quinlan, R. C4.5: Programs for Machine Learning. Morgan Kaufmann Publishers, San Mateo (1993)
22. Aha, D., Kibler, D.: Instance-based learning algorithms. Mach. Learn. **6**, 37–66 (1991)
23. Breiman, L.: Random forests. Mach. Learn. **45**(1), 5–32 (2001)
24. Holte, R.C.: Very simple classification rules perform well on most commonly used datasets. Mach. Learn. **11**, 63–91 (1993)
25. Kohavi, R.: The power of decision tables. In: 8th European Conference on Machine Learning, 174–189 (1995)
26. Berthold, M.R., Cebron, N., Dill, F., Gabriel, T.R., Kötter, T., Meinl, T., Ohl, P., Sieb, C., Thiel, K., Wiswedel, B.: KNIME: The Konstanz Information Miner, Studies in Classification, Data Analysis, and Knowledge Organization (GfKL 2007). Springer, Berlin (2007)
27. Hall, M., Frank, E., Holmes, G., Pfahringer, B., Reutemann, P., Witten, I.H.: The WEKA data mining software: an update. SIGKDD Explor. **11**(1), 10–18 (2009)

Light-Weighted DNA-Based Cryptographic Mechanism Against Chosen Cipher Text Attacks

E. Suresh Babu, C. Nagaraju and M.H.M. Krishna Prasad

Abstract DNA cryptography is a new cryptographic paradigm from hastily growing biomolecular computation, as its computational power will determine next generation computing. As technology is growing much faster, data protection is getting more important and it is necessary to design the unbreakable encryption technology to protect the information. In this paper, we proposed a biotic DNA-based secret key cryptographic mechanism, seeing as DNA computing had made great strides in ultracompact information storage, vast parallelism, and exceptional energy efficiency. This Biotic Pseudo DNA cryptography method is based upon the genetic information on biological systems. This method makes use of splicing system to improve security and random multiple key sequence to increase the degree of diffusion and confusion, which makes resulting cipher texts difficult to decipher and makes to realize a perfect secrecy system. Moreover, we also modeled the DNA-assembled public key cryptography for effective storage of public key as well as double binded encryption scheme for a given message. The formal and experimental analysis not only shows that this method is powerful against brute force attack and chosen cipher text attacks, but also it is very efficient in storage, computation as well as transmission.

Keywords DNA-based symmetric cryptography · Brute force attack · Chosen cipher text attack

E. Suresh Babu (✉)
JNTUK, & K L University, Hyderabad, AP, India
e-mail: sureshbabu.erukala@gmail.com

C. Nagaraju
Department of CSE, YV University, Proddutur, AP, India
e-mail: nagaraju.c@gmail.com

M.H.M. Krishna Prasad
Department of CSE UEC, JNTUK, Kakinada, AP, India
e-mail: krishnaprasad.mhm@gmail.com

© Springer India 2016   123
R. Chaki et al. (eds.), *Advanced Computing and Systems for Security*,
Advances in Intelligent Systems and Computing 395,
DOI 10.1007/978-81-322-2650-5_9

1 Introduction

DNA computing is a biomolecular computation (BMC), which makes use of biological methods for performing massively parallel computation. As power of the parallel processing is increasing day to day, modern cryptosystems can be easily cryptanalyzed by the cryptanalyst, the world is looking for new ways of information and network security in order to safeguard the data as it carries. The purpose of using cryptography in the areas of biomolecular computation is to bring up a promising technology for providing unbreakable algorithms. DNA cryptography is a new cryptographic paradigm from hastily growing biomolecular computation, in which its computational power will determine next generation computing. As internet technology is growing much faster, which permits the users to access the intellectual property that is being transferred over the internet can be easily acquired and is vulnerable to many security attacks like wormhole attack, IP spoofing, black hole attack, man in the middle attack, etc. Subsequently, securing all the information passed through networked computers is primarily more important for any application or system, already a great heap of effort had been put on the cryptology's, as a result, various security mechanisms have been designed such as DES, RSA, ECC, and DSA, to achieve very high level of security. However, these mechanisms require complex factorization of large prime numbers and the elliptic curve problem, for which still a lot of investigation is required to find a proper solution. Moreover, the RSA cryptosystem is based on the intractability of large prime factorization, as there are no known efficient algorithms to find largest prime factors. Hence, there is a necessity to design the unbreakable light-weighted cryptosystem to protect the information. This paper addresses a biotic DNA-based secret key cryptographic mechanism against chosen cipher attacks. Moreover, this DNA-based cryptographic mechanism makes use of DNA computation that has great strides of ultracompact information storage, vast parallelism, and exceptional energy efficiency. More importantly, this pseudo DNA-based cryptographic technique is based on central dogmas (genetic information) of biological system, similarly which is not same as original DNA-based cryptography [1–3]. In other words, this proposed method only makes use of DNA mechanisms and terminology of DNA function rather than actual biological DNA sequences (or oligos). To be more specific, this technique makes use of splicing system and multiple random key sequence to increase the degree of diffusion and confusion that provides resulting cipher texts difficult to decipher and finally makes to realize a flawless secrecy system. The experimental analysis not only shows that this method is powerful against brute force attack and chosen cipher text attacks, but also efficient in storage, computation as well as transmission.

The remainder of the paper is organized as follows. In Sect. 2 specifies the related work. Sections 3 and 4 describe the scope of research and overview of DNA. The proposed Pseudo DNA-based symmetric cryptosystem mechanism and its security analysis are discussed in Sects. 5 and 6. Section 7 describes the simulation results. Finally, Sect. 8 concludes this paper with future work.

2 Related Work

The domain of information and network security is persistently looking for unbreakable cryptosystem to protect the information while transmitting on to the network, but it seems that every cryptographic encryption technology comes across its end game as the new computing technologies are evolving.

DNA is very potent and exciting study direction from a cryptographic point of view which requires simple and effective algorithms, of late, many scientists have projected a various DNA-based encryption algorithms, but it is too early to decide the perfect complete model for some cryptographic functions, such as DNA authentication methods, digital signature, and secure data storage, as these cryptographic models are still in the initial phase. Adleman [4, 5] proposed the hypothetical model of DNA computing for any biomolecular computational problem which provides vast parallel computing. As his background stemmed from computer encryption, he particularly envisioned DNA computing in helping to create encryption and decryption algorithms in the area of cryptography. Gehani et al. from Duke University had investigated the procedure of DNA-based cryptography [3] for one-time pads (OTP). They proposed the large number short sequence of message can be encrypted using one-time pads. These small sequences of DNA can avail from massive one-time pad using public key infrastructure (PKI) [6]; Leier [2] proposed the data hiding procedure predicated on DNA binary sequences to achieve DNA encryption scheme. Applying DNA computation, Kazuo [7] resolved the trouble for generation of key distribution, he also proposed DNA-based secret key encryption system. Amin [8] proposed DNA YAEA encryption algorithm which is a conventional secret key encryption algorithm. Ning [9] proposed pseudo DNA-based cryptography along with the initial secret key to build DNA cryptosystem which is also a symmetric encryption algorithm.

3 Scope of Research Work

As power of the parallel processing is increasing day to day, modern cryptosystems can be easily cryptanalyzed by the cryptanalyst, the world is looking for new ways of information and network security in order to safeguard the data as it carries. The purpose of using cryptography in the areas of biomolecular computation is to bring up a promising technology for providing of unbreakable algorithms. However, this DNA cryptography lacks the related theory, which is nevertheless still an open problem to model the good DNA cryptographic schemes.

Recently, researchers have been devised to break the cryptosystem using DNA-based cryptography technique. In [4, 10, 11] proposed a self-assembly of DNA tiles techniques, which is used to fully break RSA scheme. If these techniques are able to break RSA, RSA will no more remain practical. Furthermore, DNA-based methods had also been developed to break the cryptosystems based on elliptic curves [12]. These methods utilize a parallel multiplier to perform basic biological operations and for adding the points on elliptic curves, it uses both parallel divider and a parallel adder. Moreover, so far many researchers had concentrated on breaking the cryptosystem using several DNA-based methods, which are presently being practiced.

4 Overview of DNA Computation

DNA computing is a BMC which makes use of biological methods for performing massively parallel computation. This can be a lot quicker than a conventional silicon chip computer, for which large quantities of hardware needed for performing parallel computation. These DNA computers [5, 13–16] not only just makes use of massively parallel computation, but also uses ultracompact information storage in which large amount of information that can be stashed in a more compact away with, which massively exceeds in conventional electronic media (i.e., a single gram of DNA [5, 17, 18] comprises 1021 DNA bases which equals to 108 terabytes). A hardly few grams of DNA possibly contain all stored data in the world. Moreover, this cross-topical field of DNA computing [19] combines the ideas from biological sciences, computer science, and chemistry. In 1994, Adleman [18] designed a study to solve the travelling salesman problem that attempts to visit each city exactly once and try to find every possible route using molecules of DNA. Hence, this inspired model provides the potential ability of working out many problems that were previously thought impossible or exceedingly difficult to solve out with the traditional computing paradigm such as encryption breaking and game strategy.

5 Background of DNA Cryptography

In order to understand the rudimentary principles of DNA cryptography in the emerging area of DNA computing, it is necessary to address the background details of central dogma of molecular biology. Indeed, deoxyribo nucleic acid (DNA) is the fundamental hereditary material that stores genetic information found in almost every living organisms ranging from very small viruses to complex human beings. It is constituted by nucleotides which form polymer chains. These chains are also known as DNA strands. Each DNA nucleotides contains a single base and usually consists of four bases, specifically, thymine (T), cytosine(C), and guanine (G) and adenine (A) represent genetic code. These bases read from the start promoter which forms the structure of DNA strand by forming two strands of hydrogen bonds, one is A with T and another is C with G; These DNA sequences are eventually transcript and interpreted into chains of amino acids, which constitute proteins, which is depicted in Fig. 1. In other words, the encryption process is initiated with DNA

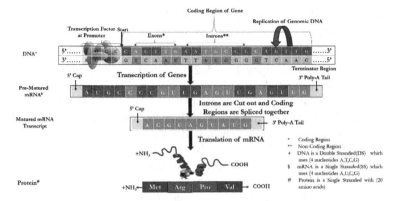

Fig. 1 Central dogma of molecular biology

transcription along with splicing system, RNA translation [9, 20] will be transformed into a protein sequence. However, the decryption process is performed in reverse order of encryption.

5.1 Transcription

Transcription is a process of newly prepared intermediary copy of DNA called mRNA instructions that transpires the DNA sequence [21, 22], which comprises of nucleotides A, G, C, and T. For instance, DNA sequences are transcript into mRNA sequence, here mRNA is a single stranded that contains the nucleotides A, G, C, and Uracil (U). Moreover, the transcript process of RNA undergoes the processing step called splicing steps, in which expressed (Extron's) are the coding regions that are spliced and intervening (introns) which are noncoding regions are cut out to form mRNA sequence. The intermediary mRNA polymerase (enzyme) which is responsible for copying DNA into RNA.

5.2 Translation

Translation is another process that contains the RNA copy of DNA to make a protein sequence. i.e., the mRNA copy of DNA sequence is translated into different amino acids that can be bound to collect protein sequence. There are twenty different amino acids, which is a basic building block of a protein sequence. Eventually, this mRNA translation process makes use of codons (collections of 3 nucleotides) [19, 23] that are translated into the amino acids, according to the genetic code table [24].

6 Modeling a Pseudo DNA-Based Symmetric Cryptosystem

The pseudo DNA cryptography [9, 25–27] method makes use of the standard principle ideas of central dogmas (genetic information) of biological system. Especially, the pseudo DNA (encryption/decryption) cryptosystem process is similar to the DNA transcription process along with splicing system and RNA translation process of the real organisms. However, it is different from existing DNA-based cryptography [28–32], as this method only make use of DNA mechanisms and terminology of DNA function rather than actual biological DNA sequences (or oligos); Hence, this proposed method is a kind of pseudo biotic DNA-based cryptography method. The pseudo DNA cryptography technique consists of transcription/splicing system and translation processes as specified central dogma of molecular biology, which is depicted in Fig. 1. Moreover, In order to make the statistics of the cipher text more difficult for the attacker to decipher, we integrated multiple rounds of random keys into the encryption algorithm by modifying the original splicing system. Originally, the starting codes of the introns and ending codes of the introns are very easy to guess. In this proposed work, we have modified start codes and the pattern codes to specify the introns. The noncontinuous pattern codes are used to confuse the adversary and hard to guess the introns, by defining which parts of the DNA frame is to be removed, and which DNA frame is to be kept. Furthermore, the no of the splices, the starting code of the frame, and removed length of the pattern codes can be used to determine the key. However, the ending codes of the DNA frame are not required in this case.

6.1 Symmetric Cryptography Principles

Generally, modern cryptography [33] solves many cryptographic algorithms with the help a KEY. The cryptosystem which comprises of encryption and decryption functions using the same key(K) that can be interpreted as symmetric cryptography, which is represented with two functions: $E_K(M) = C$ and $D_K(C) = M$. In this cryptosystem, first, both the sender and receiver must agree on a key as well as cryptosystem in order to communicate securely. Hence, the success of such symmetric cryptosystem is mainly depends upon its Key.

6.2 Communications Using Pseudo DNA Symmetric cryptography

The conventional secret key encryption scheme $\Pi = (\mathcal{E}_\mathcal{K}, \mathcal{D}_\mathcal{K})$ is usually represented with two algorithms; one is $\mathcal{E}_\mathcal{K}$ function which is a stateful encryption algorithm with k randomized key generation algorithm. It takes the plaintext 'M' along with random key 'K' and returns a cipher text 'C'; usually represented $E_K(M) = C$ and another is $\mathcal{D}_{\mathcal{K}}$ function which is a deterministic decryption algorithm, which takes a string 'C' and the same random key 'K' and returns the equivalent plaintext 'M'; usually represented as $D_K(C) = M$ where $M \in \{A, T, C, G\}^*$; finally we perform that $D_k(E_k(M)) = M$ or all $M \in \{A, T, C, G, 0, 1\}^*$

In order to perform above communications model using symmetric pseudo DNA-based cryptography, the following steps can be described briefly as shown in Fig. 2

1. Alice takes the plaintext and it converts into binary form which in turn converts into DNA form
2. Alice will scan DNA form of information to generate the variable length random key by generating the no of the splices from the specified DNA pattern, the starting code of the DNA frame to find out the introns, introns places, and removed length of the pattern codes i.e., introns are removed from the specified DNA sequence as the first round of key generation
3. With the help of random key (splicing system), Alice will transcript the DNA sequence into the mRNA strand
4. After Generating mRNA strand, Alice also generate the variable length random subkey by generating the no of the splices from the specified mRNA pattern, the starting code of the mRNA frame, introns places, and removed length of the pattern codes as the second round of processing as depicted in Fig. 2

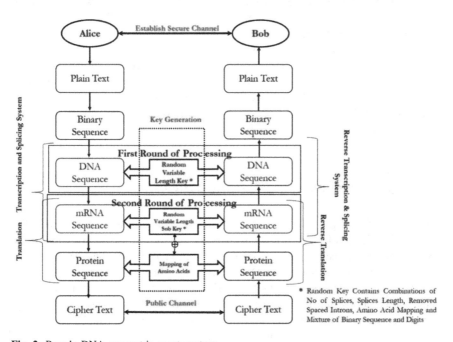

Fig. 2 Pseudo DNA symmetric cryptosystem

5. Again, the spliced mRNA strand is translated into the amino acids, which forms protein sequence, as shown in Algorithm-5 (Hint: The mapping of codons to amino acids is done with the help of genetic code table. Usually table consists of 61 codons, which are mapped 20 amino acids)

6. Next, the protein sequence (Cipher Text) will be sent to the Bob through public channel.

7. The random variable length key is comprised number of splices, the starting index, pattern codes length of the introns, the positions, and places of the introns, and the cut out the introns; random mapping of codon-amino acids will form the symmetric key to decrypt the cipher text (protein sequence), which is also sent to the Bob through a secure channel as shown in Fig. 2.

8. On Bob's (Receiver) side, when he receives random keys and protein form (Cipher text) of data from Alice through the secure channel, then he can perform the decryption process.

9. Bob decrypts the cipher text message using the random key reversible translation to recover mRNA sequence from protein sequence, and then recover DNA form of information, in the reverse order as Alice encrypt the information.

10. Bob can then recover then binary form of information and finally get what Alice sent him.

The biotic DNA symmetric cryptosystem is designed in such way that the adversary cannot decrypt the encryption algorithm without the information of the key; Moreover, it is very difficult to find the random DNA secret key sequence and random mRNA key sequence and random mapping of amino acids. To be more specific, suppose, if the adversary applies brute force search for finding the random key in order to decrypt the cipher text, then the attacker should spend numerous time and resources. Because, DNA is known for enormously huge quantity of data storage, which requires heaps of nucleotides to find the correct no of splices, cutoff introns, starting position of DNA/mRNA strand, removed DNA/mRNA strand and mapping of Amino acids. Hence, the algorithm is secure and safe enough against Brute force attack and chosen cipher text (CCA) [34], which will be discussed in next subsequent section

6.3   Key Generation Using Splicing Systems

Tom Head [35, 36] proposed the splicing system which captures mathematically $\Sigma_{DNA} = \{A, C, G, T\}$. Where DNA strands are referred as strings over the finite alphabet. However, these splicing systems were introduced more than 20 years ago, that is when nobody spoke about DNA computing. In fact, only in 1995—thus, after Adleman's paper—splicing systems [37, 38] have been suggested to represent DNA computations and their computational properties, by various authors. The central operation of the splicing systems: given an alphabet S and two strings, $y \in \Sigma^*$, it is defined the splicing of x and y, as indicated by the rule r. Formally, a splicing rule r defined on the alphabet Σ is a word of the form $\alpha_1 \# \beta_1 \$ \alpha_2 \# \beta_2$, where $\alpha_1, \beta_1, \alpha_2, \beta_2 \in \Sigma^*$, while # and $ are special symbols that are part of Σ. If we have $x = x_1 \alpha_1, \beta_1, \alpha_1' \ y = y_2 \alpha_2, \beta_2, y_2'$ and $r = \alpha_1 \# \beta_1 \$ \alpha_2 \# \beta_2$. we write: $(x, y) \rightarrow_r (p, q)$ to indicate that the strings p and q are obtained from the values of x and y applying the splicing rule r.

6.3.1 Key Selection Using Splicing Systems

The security of any cryptosystem relies on key management. This proposed algorithm improves the security levels that make use of random key generator, which is based upon splicing system of central dogma. The key information will be selected from DNA sequence, which can be generated randomly by the user. Similarly, the subkey information will also be selected from mRNA sequence, which is generated randomly by the pseudo random generator algorithm. Subsequently, these two random keys will be XORed with random mapping of codon-amino acids. Finally, resultant random key will be shared between Alice and Bob through private or secure channel as shown in Fig. 2.

7 Modeling an Application (Public Key Cryptography) Using Pseudo DNA Symmetric Cryptosystem

Definition of Public Key Cryptography: An Asymmetric Cryptosystem is mathematical function with three-tuple $\prod = (\mathcal{E}, \mathcal{D}, \mathcal{K})$ where $\mathcal{E}$ is the encryption algorithm and $\mathcal{D}$ is the Decryption algorithm and $\mathcal{K}$ is a key generation algorithm. It returns two keys in which one is a DNA public key denoted by K(Dpub, s) and another DNA private key denoted by K(Dpri, s); we write $a_1 \leftarrow K_{(Dpub,s)}$ and $a_2 \leftarrow K_{(Dpri,s)}$; $\mathcal{E}(\mathcal{M}, \mathcal{K})$ is a stateful public key algorithm. The deterministic encryption algorithm takes a plaintext 'p' along with the key 'a1' and returns a cipher text 'q'; we denote as $q \leftarrow E_a(p)$; $\mathcal{D}(C, K)$ is a decryption algorithm which takes the cipher text 'q' along with DNA private key 'a2' and returns the equivalent plaintext 'p'; we denote as $p \leftarrow D_a(q)$, where $p \epsilon \{0, 1, A, G, T, C\}^*$; finally we represent the stateful public key algorithm with $D_{a2}(E_{a1}(p)) = p$ for all $p \epsilon \{0, 1, A, G, T, C\}^*$.

We are using underlying RSA algorithm for encrypting the plain text message, which is one of the *public key cryptosystem*. Before we discuss the working process of DNA-based *public key cryptosystem*, let us outline and give a brief definition of RSA algorithm. According to the definition, first step is to generate the key. To achieve this, initially, choose the two largest and distinct prime numbers say p, q. Usually these numbers should be in the order 10^{100} which makes the algorithm very robust. Next step is to calculate $N = P \times Q$ to get the product. Here N is the product of PQ. As RSA is known for public key and private key usage, therefore, next step to find the integer 'e' is a public key and 'd' is the private key. Furthermore, to calculate 'e' and 'd'. Indeed, $e.d \equiv ((\bmod()p - 1).(q - 1))$ then set the private key (d, N) and the public key (e, N). After the key generation, next phase is encryption algorithms, whose output is usually represented as $E_{\text{RSA}}(m, (e, N)) \doteq m^e \bmod N = c$, where E_{RSA} is RSA encryption algorithm; m is the plain text message; (e, N) is the public key and c is the cipher text. Final step is decryption algorithm whose output is usually represented as $D_{\text{RSA}}(c, (d, N)) \doteq c^d \bmod N = m$; here (d, N) is the private key used for decrypting the cipher text.

7.1 DNA-Assembled Public Key Cryptography

A DNA-assembled public key encryption scheme makes use of asymmetric encryption scheme for effective storage of cipher text and public key in the DNA strand. For better understanding the communications of DNA-based symmetric cryptography. Let us illustrate with an example as shown in Fig. 3

1. Let us assume the Alice wants to send the message to the Bob;
2. Both of them should agree on an asymmetric cryptosystem.
3. Next, Alice takes her plaintext message and encrypts the plain text using RSA algorithm (procedure discussed above), which in turn generates cipher text and the public key. Notice that cipher text is comprises of numerical values. For instance the original message is "welcome to kl university" and after performing RSA encryption we get cipher text "1191011089911110910111161111071 0811711010511810111411151 05116121" public key is "7E823AF77CA013E AEDDB118077EF1A39EA0538E" .

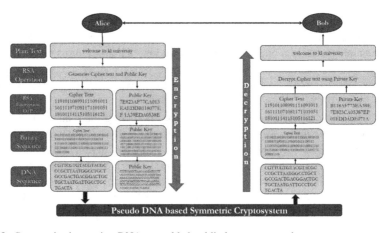

Fig. 3 Communications using DNA-assembled public key cryptography

Table 1 Comparison of key length of public key length and symmetric key length versus DNA-based symmetric key length and DNA-based symmetric key length

Public key length	Symmetric key length	DNA-based symmetric key length	DNA-based symmetric key length
56 bits	384 bits	14 bits	96 bits
64 bits	512 bits	16 bits	128 bits
80 bits	768 bits	20 bits	192 bits
112 bits	1792 bits	28 bits	448 bits
128 bits	2304 bits	32 bits	576 bits

4. Once again, Alice converts both the cipher text and public key into binary form. For instance cipher text equivalent binary string "0110111101101100011 0110001100 101011001111000011101001011110011111100101100000111100 001101000111100111000011100011111100101111001111000011100" and public key equivalent binary string "0111011101100100110110000100000 01100011011011110110110 01011001010010000011101000110111110010 00100000111000001100001011000010110111101101000110110001100101011 001110110010100100000011000110110110011011001010010000000110010001110 01011100001101100101011011100111010000000101."

5. Next, she transforms both the binary forms into DNA form (A for 00, C for 01, G for 10, T for 11) cipher text equivalent DNA form "CGTTCGTGTACG TACGCC CGCTTAATGGCCTGCTTGCCGACTGACGGACTGCTGCTAA TGATTGCCTGCTGACTA" and public key equivalent DNA form "CTCT CGCCCGTAA GAACGATCGTTCGTCCCGCCAGAACTCACGTTAGAGA ATGAATAAGTACCGTTCGTACGTACGCCCGCTCGCAGAACGATCTAT CGCCAGAACGCACGCCTAATCGCCCGTGCTCAACC."

6. Finally, Alice takes the DNA form and performs pseudo DNA-based symmetric cryptography(discussed in section) that convert the cipher text message and sends to Bob through the network;

The main advantages of this method, first, it provides double encryption, which is very difficult for adversary to break the cryptosystem. Second, it provides more compact storage space, which is more suitable for public key encryption. In our case, RSA public keys require huge storage for storing large prime numbers, while DNA-assembled public key cryptography takes less storage than public encryption algorithm as shown in Table 1.

8 Security Analysis

The main objective of any security technique is to strengthen and to protect the network and information from any malicious activities. Subsequently, time and computational complexity are the two important parameters need to compute for any sort of cryptographic schemes. Particularly, semantic security and message indistinguishability are the two fundamental computational complexity analog of Shannon's definition of perfect privacy [39]. Former one represents the infeasibility to learn anything about the plaintext from the cipher text and the later one represents the infeasibility of distinguishing between the given pair of messages.

8.1 Formal Definitions of Semantic Security (SS) and Message Indistinguishability (MI)

Definition 1 Semantic security ensures that nothing can be learned just by looking at a cipher text. i.e., cipher text reveals no information about the message. For every distribution X over $\{0, 1\}^n$ and for every partial information $h:\{0, 1\}^n \rightarrow \{0, 1\}^n$ for every interesting information $f:\{0, 1\}^n \rightarrow \{0, 1\}^n$. For every attacking algorithm A, running time complexity $t' \leq t(n)$ [$t(n)$ is a polynomial in n], \there exists algorithm S such that:

$$\Pr_{m \leftarrow X, (P_k, S_k) \leftarrow G(n)}[A(E(m, p_k), p_k, h(m)) = f(m)] \leq \Pr_{m \leftarrow X}[S(h(m)) = f(m)] + \varepsilon(n),$$
(1)

where $\varepsilon(n)$ is a negligible quantity which depends upon value of n. Ex. $\varepsilon(n)$ may be $\frac{1}{P(n)}$, where $p(n)$ is a polynomial in 'n' of a large degree.

Definition 2 Given two encryptions of messages m_0 and m_1, the probability of guessing the message is very close to the random probability of guessing the correct message $(\frac{1}{2})$. The security of message indistinguishability states that the inability to distinguish two plaintexts (of the same length) i.e., the cipher texts are computationally indistinguishable. For every two messages m_0, $m_1 \in \{0, 1\}^n$; for every algorithm A that runs within time $\leq t(n)$

$$\Pr_{\substack{i \in \{0,1\}, \\ (P_k, S_k) \leftarrow G(n)}} [A(E(m_i, p_k), p_k) = i] \leq] \leq \frac{1}{2} + \epsilon(n)$$
(2)

The theoretical result of symmetric DNA-based encryption function gives a diffusion cipher text, which is hard to compute plaintext without random key. Therefore, security analysis of symmetric DNA-based cryptography is efficient and very powerful against certain cryptographic attacks.

Definition 3 A polynomial time-computable predicate b is called a hard-core of a function f, if every efficient algorithm, given $f(x)$, can guess $b(x)$ with success probability that is only negligibly better than one-half. Formally speaking, we define a hard-core predicate as follows: a polynomial time-computable predicate $b : \{0,1,G,T,A,C\}^* \rightarrow \{0,1,G,T,A,C\}$ is called a hard-core of a function f if for every probabilistic polynomial time algorithm A', every positive polynomial $p(.)$, and all sufficiently large n's,

$$P_r\left[A'(f(U_n)) = b(U_n)\right] < \frac{1}{2} + \frac{1}{p(n)}. \tag{3}$$

Note that, for every $b : \{0,1,G,T,A,C\}^* \rightarrow \{0,1,G,T,A,C\}$ and $f : \{G,T,A,C\}^* \rightarrow \{G,T,A,C\}$, there exist obvious algorithms that guess $b(U_n)$ from $f(U_n)$ with success probability at least one-half, e.g., the algorithm that obliviously of its input outputs uniformly chosen DNA strand. Also if b is a hard-core predicate for any function, then $b(U_n)$ must be almost unbiased (i.e., $|P_r[b(U_n) = 0] - [b(U_n) = 1]$ must be a negligible function of n). Now our encryption scheme make use hard-core predicate (hp) and we analyze the security of the scheme.

8.2 Encryption Algorithm

Assume the encryption function (Fbin, Fdna Frna, Fpro) and a hard-core predicate B(X,k) for FKEY. Here we want to encrypt a plaintext p and b is a key, which is the secret information.

"**Theorem:** Symmetric DNA based encryption scheme for Message i.e. Encryption EFbin, Fdna Frna, Fpro (b, k) is MI secure.

1. SCHEME ((Fbin, Fdna Frna, Fpro ,FKey),B,b)
    /*** Encryption EFbin, Fdna Frna, Fpro (b, k) ***/

    pick $X \overset{U}{\leftarrow} \{G,T,A,C\}^n$

    Return $\left(F(X,k), b, B(X, k)\right)$

2. /*** key generation ***/
    Generate the Combination of pairs (kb, Kd, Kdna) using Fdna and Rdna

3. /*** Decryption DFbin, Fdna Frna, Fpro (c, F(X,k)) ***/
    $X = D[F(X,k), K_b, K_d, K_{dna}]$

    Return $(c, B(X,k))$

Proof : Suppose the encryption scheme is not $(t, \in)$-MI secure, So $\exists$ a PPT algorithm A' such that

$$\Pr_{\substack{b \in \{0,1,G,T,A,C\}, \\ (P_k,S_k) \leftarrow G(n) \\ X \overset{U}{\leftarrow} \{G,T,A,C\}^n}} \left[A\left(F(X, k), b, B(X, k), k \right) = B \right] \leq \frac{1}{2} + \in (n)$$

Consider the following algorithm A'

A' (y, k)

{

    1. Pick random $c \in$ {protein form}

    2. Return $\left(c, A\left(y, c, k \right) \right)$

}

So,

$$\Pr_{X \overset{U}{\leftarrow} \{G,T,A,C\}^n} \left[A'\left(F(X, k), b, B(X, k), k \right) \right]$$

$$= \Pr_{\substack{c \in \{\text{proteins form}\}, \\ (P_k,S_k) \leftarrow G(n) \\ X \overset{U}{\leftarrow} \{G,T,A,C\}^n}} \left[A\left(F(X, k), c, B(X, k), c \right) \right] > \frac{1}{2} + \in (n)$$

Since A' is a PPT algorithm just as A. So B is not a hard-core predicate (hp) according to definition. This is a contradiction. Hence, the primary assumption was wrong. Hence SCHEME ((Fbin, Fdna Frna, Fpro),B,b) FKey is MI secure. Hence proved"

9   Cipher Text Indistinguishability

Cipher text indistinguishability is a one of the important security property for numerous encryption schemes. Instinctively, if a cryptosystem has the property of indistinguishability, then an opponent will be not able to distinguish pairs of cipher texts focused around the message they encrypt. The property of indistinguishability is viewed as an essential requirement for most of the provably secure key cryptosystems under chosen cipher text attack, chosen plaintext attack, and adaptive chosen cipher text attack. A cryptosystem is viewed as "secure in terms of indistinguishability" if no opponent A, given an encryption of a message haphazardly chosen from a two-component message space controlled by the opponent, can distinguish the message decision with likelihood better than that of random guessing (1/2). If any opponent can succeed in recognizing the chosen cipher text with likelihood fundamentally more noteworthy than ½. There are numerous security definitions in terms of indistinguishability, depending on presumptions made about the abilities of the attacker. At this point, when the cryptosystem is viewed as secure, if no opponent can guess randomly with significantly probability more prominent better than half. The most well-known definitions used in cryptography are indistinguishability with various attacks [34] such as (nonadaptive) chosen cipher text attack (IND-CCA), chosen plaintext attack (IND-CPA), adaptive chosen cipher text attack (IND-CCA2). The convenient way to sort out above definitions to secure DNA-based Encryption is by considering different conceivable objectives and attacks models. The objective here is to make an opponent's powerlessness to realize any data about plaintext underlying a challenge cipher text. In this conception, the adversary cannot determine from which plaintext the challenge cipher text came from.

The attack models are considered here are adaptive chosen cipher text attack, nonadaptive chosen cipher text attack, and chosen Plain text Attack (CPA). In IND-CPA is characterized between an opponent and a challenger. For schemes focused around computational security, the adversary is modeled in such a way; he must finish inside a polynomial number of time steps to guess. In IND-CCA1, the adversary has a right to access to unscrambling oracle O. Nevertheless, the opponent can utilize this oracle only before it gets the challenge cipher text y. Finally, In IND-CCA2, adversary has a right to gain the access of oracle O and his inquiry to the oracle may rely on upon the challenge cipher text y. However, the only restriction with this attack is that the opponent can not query the oracle to the challenger to decrypt the cipher text y . In formalizing IND-Atk, an opponent 'A' as a pair of probabilistic polynomial time algorithm A = (A_1, A_2). Here, A runs in two stages. Whereas A1 generates a message pair, encrypts one of them, and sends to A2 as challenge cipher text. We say A2 is successful depending on its goal; the goal is here to tell which message is in encrypted form.

9.1 Indistinguishability of IND-CCA1 or IND-CCA2

Definition 4 Let $\prod = \{\mathcal{E}, \mathcal{D}, \mathcal{K}\}$ be a secret key encryption scheme. For an opponent A and $b = \{0;1\}$ characterize the experiment
Experiment:

$$\text{Exp}_{\pi}^{\text{ind}-\text{cca}}(A, b)$$
$$a \leftarrow K; (x_1, x_2, s) \leftarrow A^{E_a D_a}(\text{Find});$$
$$y \leftarrow E_a(x_b); d \leftarrow A^{E_a D_a}(\text{Guess}, y, s)$$
$$\text{Return } d;$$

It is assigned that $|x_0| = |x_1|$ above and that opponent A does not query for decryption oracle $D_a(\cdot)$ on cipher text y in the supposition phase. Characterize the advantage between opponent A and function π respectfully, takes as follows:

$$\text{Adv}_\pi^{\text{ind}-\text{cca}}(A) = P_r\left[\text{Exp}_\pi^{\text{ind}-\text{cca}}(A,0) = 0\right] - P_r\left[\text{Exp}_\pi^{\text{ind}-\text{cca}}(A,1) = 1\right] \qquad (4)$$

$$\text{Adv}_\pi^{\text{ind}-\text{cca}}(t, q_e, q_d, \mu, v) = \overset{\text{MAX}}{A} \text{Adv}_\pi^{\text{ind}-\text{cca}}(A) \qquad (5)$$

The maximum time complexity t with at most q_e and q_d encryption and decryption oracle queries and totaling these queries with at most μ bits and finally choosing $|x_0| = |x_1| = v$ bits. Hence, the worst-case time complexity for this experiment is $\text{Exp}_\pi^{\text{ind}-\text{cca}}(A)$ plus the total size of the code of opponent A."

The analogy of the above definition E (P_K, "M") which represents the encrypted message "M" under the random key "P_K": The challenger produces encrypts arbitrary cipher texts, and the opponent is offered to access the decryption oracle, which decrypts self-assertive cipher texts at the opponent's request, retaining the plaintext. The opponent may keep on query the decryption oracle significantly even after it has received a challenge cipher text, but it may not pass the cipher text for further processing

Step-1: The challenger generates a key "P_K" in multiple rounds of transcription (first key), spicing system (second key) and translation process (third key) (e.g., a key size in Kdna, Kmrna, Kmap) which produces cipher text and given to the opponent

Step-2: The opponent calls to the decryption function based on haphazard cipher texts

Step-3: The challenger selects the key P_k = {Kb, Kd, Kpdna} randomly and sends the challenge cipher text $\mathbb{C} = E(P_k, M)$ back to the opponent

Step-4: The opponent is free to execute any number of encryptions or computations

Step-5: Once again, the opponent may further calls to the decryption function, but this time he may not submit the cipher text "C"

Step-6: Finally, the opponent generates the outputs by guessing for the value of message "M". This scheme is secure against IND-CCA2 if no opponent can guess with nonnegligible time

A DNA-based private key scheme ((Fbin, Fdna Frna, Fpro,FKey),B, b) is (t, q, ϵ) secure in IND-Atk sense. If for all pair of different messages of same length and any opponent A that runs within given time t and performs at most q queries to the decryption oracle O, $\epsilon(n)$ denotes the advantage of the algorithm over a random guess.

$$Pr_{(P_k,S_k)\leftarrow G(n)}\left[A^0\left(P_k E_{pk}(m_1)\right) = 1\right] - Pr_{(P_k,S_k)\leftarrow G(n)}\left[A^0\left(P_k E_{pk}(m_0)\right) = 1\right] \leq \epsilon(n),$$
$$(6)$$

where the oracle is

$$O = \begin{cases} - & \text{if IND} - \text{CPA} \\ D_{sk} & \text{if IND} - \text{CCA2} \end{cases}$$

and the adversary cannot query the decryption oracle at EP_k (m_i). Therefore, informally an pseudo DNA-based encryption scheme is secure if for each adversary A and for every polynomial $P(.)$, there exist a 'N' such that

$$(A \text{ succeeds in the attack}) < \frac{1}{P(n)} \forall n > N. \tag{7}$$

From the definition of semantic security, for all distributions over $\{A,T,G,C\}$; for All Partial informations $h:\{\text{Proteins}\}^n \to \{\text{Proteins}\}^n$; for all interesting informations $f : \{0,1,A,T,C,G\} \to \{O, 1, \text{digits}, \text{DNA Strands}\}$; adversary A with time complexity $t' < t(n), t(n) = \sum t_d n^d$; $\exists$ simulating algorithm S such that

$$P_{\substack{r(P_k,S_k) \leftarrow G(n) \\ X \xleftarrow{U} \{G,T,A,C\}^n}} [A(E(m, p_k), p_k, h(m))) = f(m)] \leq$$

$$P_{\substack{r(P_k,S_k) \leftarrow G(n) \\ X \xleftarrow{U} \{0,1,G,T,A,C\}^n}} [S(h(m)) = f(m)] + \varepsilon(n) \tag{8}$$

where "$\varepsilon(n)$ is a negligible quantity; then $E(.)$ is called semantically (t, ε)" is secure
From the definition of message indistinguishability; for all messages $m_0 m_1 \in \{0,1\}^n$; for all adversary A with time complexity $t' < t(n), t(n) = \sum t_d n^d$

$$Pr_{\substack{i \in \{0,1\}, \\ (P_k, S_k) \leftarrow G(n)}} [A(E(m_i, p_k), p_k)) = i] \leq \frac{1}{2} + \in (n), \tag{9}$$

where $\varepsilon(n)$ is a negligible quantity: then $E(.)$ is called $(t; \varepsilon)$ MI secure; n is the security parameter such as key length; $\varepsilon(n)$ is a negligible quantity.

10   Results and Simulations Analysis

To study the feasibility of our theoretical work, we have implemented and evaluated the pseudo Biotic DNA cryptography method in C++ and conducted a series of experiments in a network simulator [NS2] to evaluate its effectiveness. The experiment results show that this method is more efficient and its increase the power against certain adaptive cryptographic attacks. The experimental values were obtained by evaluating the multiple running times of the pseudo biotic DNA cryptography on a software program running Uduntu-13.04. Our simulations are based on sender and receiver programs. On the sender side, the sender first converts the plaintext into the binary sequence, which in turn translated into the DNA strand. Indeed, necessary padding is performed at the time of translation in order to have the compatibility DNA strand. After translation, the sender will generate the random variable length key using the splicing system process of the central dogma. In other words, the sender will generate the random key with a mixture of binary sequence, decimal digit, and DNA Strand, which makes the adversary hard to guess the key and translates into mRNA sequence. Next, the subkey generation is chosen at mRNA sequence using pseudo random key generator. Subsequently, these two random keys will be XORed with random mapping of codon-amino acids to produce the protein sequence. To put in another way, the mRNA is translated into the amino acid sequence called codons, which produces the proteins sequence. Eventually, the whole transcription and translation process of central dogma creates enciphered information. These enciphered information and random key are transferred to receiver through different channels i.e., enciphered information through public channel, and random key through secure channel.

On the destination side, the receiver receives the enciphered information and random key from different channels. Consequently, the receiver uses decryption algorithm and the same key information to decipher the enciphered information. To be more specific, first, the receiver performs reverse translation process to recover from protein sequence into mRNA form using same subkey with the help of pseudo random generator. Next, reverse transcription process is performed using reverse splicing process to recover from mRNA to DNA form. Finally, he recovers the plaintext using the recovery translator that the sender had sent him.

Let us exhibit the proposed biotic cryptography method with an example; Alice creates a cipher text and public key converts into the DNA Strand. "CGTTCGTGTAC GTACGCCCGCTTAATGGCCTGCTTGCCGACTGACGGA CTGCTGCTAATGATTGCCTGCTGACTA" and "CTCTCGCCCGTAAGAACG ATCGTTCGTCCCGCCAGAACTCAC GTTAGAGAATGAATAAGTACCGTTC GCCCGCTCGCAGAACGATCTATCGCCAGAACGCACATCGCCCGTGCTCA ACC," respectively. Moreover, she also generates the variable length random key (splicing system) "00011001003CGT011113TGC101113GAC" from DNA Strand of cipher text. However, DNA form of public key will be converted into equivalent numerical form for clear understanding of the key. The main specific reason of converting the public key into DNA form is to have optimized key size. Subsequently, the subkey "00011000113CTG011013CTT10 1013GAT" is chosen from mRNA sequence "GUAGGUAAUGAUCUGCUUCAUC UUGC UUGCCGACUGACGG AUUA" using pseudo random key generator. Finally, these mRNA Strand is translate into amino acid sequence (codons), which produces proteins sequence "ValGlyAsnAspLevHisLevAlaCysArgLevThrAspTALev." This encoded proteins sequence will be sent to the Bob. Bob decrypts the cipher text using the same random key to recover the plain text. We verified experimentally that the encryption and decryption can be performed effectively a given key. Moreover, different plaintexts with the combination of alphabets, digits, and few special characters are chosen with increasing size that includes short-text, average-text and long-text. Indeed, each plaintext is stored in ASCII format and number of bits are calculated to that of 8 or 16 times that of the length of the plaintext. The original plaintext size is calculated with different 64, 128, 256, 512, 1024, and 2048 bits random key and the resulting cipher text size are examined. These random key are used to examine the efficiency of the algorithm in terms of computation, storage, and transmission. Furthermore, we also investigated that the proposed algorithm needs the 264, 310, 410 and 575 chosen cipher texts to find the message without key for different key size.

As shown in Fig. 4. The length of cipher texts is proportional to that of the corresponding plaintexts lengths with varying key length. However, this method requires less storage space than that of the plaintext, thus, it is more efficient in the storage capacity. Another reflection is that the size of the random key length increase as the size of the plaintext increase, which greatly reduces size of the key length. Moreover, key as well cipher text can be transmitted much faster through the secure channel and public channel, respectively. Therefore, the method is also more efficient interms of storage and transmission.

As shown in the above Fig. 5. The adversary requires more than 65 % of chosen cipher texts for the corresponding plaintexts to recover 78 % of the random key length. Hence, it requires more chosen cipher text to retrieve the key. Figure also shows that different tests are performed to experiment the robustness of this proposed method. Therefore, it is more efficient and effective method.

The above Fig. 6 indicates, for the same plaintext length, it generates different cipher text, namely cipher text-1 and cipher text-2 with different random key. Thus,

Fig. 4 Performance analysis between plaintext, cipher text, and key length

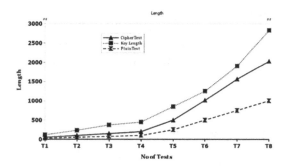

Fig. 5 Performance analysis between plaintext, chosen cipher text, and its deduction of key

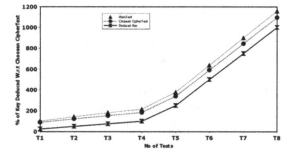

Fig. 6 Analysis of message
indistinguishability (MI) of
plaintext cipher text with
different random keys

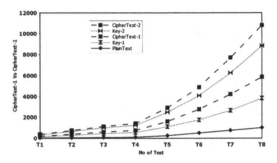

Fig. 7 Percentage of chosen
cipher text w.r.t. PPT
algorithm

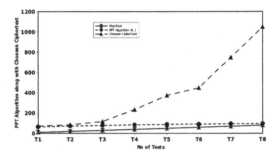

this method satisfies the message indistinguishability (MI) because the probability
of guessing these two cipher text is more than half of the random probability of
guessing the right message.

The above Fig. 7 shows that the adversary requires more chosen cipher text for a
given plaintext, which takes more than half of the time to retrieve the key.
Therefore, PPT algorithm satisfies message indistinguishability (MI), according to
the definition.

11 Conclusion

In this paper, we addressed a biotic DNA-based secret key cryptographic mechanism, which is based upon the genetic information of biological system. Moreover,
this cryptographic prototype is motivated from biomolecular computation, which is
rapidly growing field that has made great strides of ultracompact information
storage, vast parallelism, and exceptional energy efficiency. Over the last two
decades, Internet technology is growing much faster, which permits the users to
access the intellectual property that is being transferred over the internet can be
easily acquired and is vulnerable to many security attacks. Hence, network security
is looking for unbreakable encryption technology to protect the data. This motivated us to propose biotic pseudo DNA cryptography method, which makes use of
splicing system to improve security and random multiple key sequence to increase

the degree of diffusion and confusion that makes resulting cipher texts difficult to decipher and to realize a secure system. Furthermore, we also modeled DNA-assembled public key cryptography for effective storage of public key as well as double encryption scheme for a given message. The formal and experimental analysis not only shows that this method is powerful against chosen cipher text attacks, but also very effective and efficient in storage, computation as well as transmission. To conclude, DNA cryptography is a new emerge area and extremely guaranteeing field, where research is possible in incredible development and improvement.

References

1. Chen, J.: A DNA-based, biomolecular cryptography design. In: IEEE international symposium on circuits and systems (ISCAS), pp. 822–825 2003
2. Leier, A., Richter, C., Banzhaf, W., et al.: Cryptography with DNA binary strands. Biosystems 5(7), 113–122 (2000)
3. Gehani, A., LaBean, T.H., Reif, J.H.: DNA-based cryptography. In Winfree, Gifford (eds.), Proceedings 5th DIMACS Workshop on DNA Based Computers, held at the Massachusetts Institute of Technology, Cambridge, MA, USA June 14–June 15 (1999). American Mathematical Society (1999). vol. 54, pp. 233–249
4. Adleman, L.M., Rothemund, P.W.K., Roweis, S., Winfree, E.: On applying molecular computation to the data encryption standard. In: Landweber, L.F., Baum, E.B. (eds.), DNA based computers. In: Proceedings of the Second DIMACS Workshop, June 10–12 (1996), DIMACS Series in Discrete Mathematics and Theoretical Computer Science, vol. 44. American Mathematical Society, pp. 31–44 (1999)
5. Adleman, L.: On constructing a molecular computer. University of California, U.S.C Draft (1995)
6. "PKI Infrastructure". Treasury Board of Canada. October 4, 2001
7. Kazuo, T., Akimitsu, O., Isao, S.: Public-key system using DNA as a oneway function for key distribution. Biosystems 81, 25–29 (2005)
8. Amin, S.T., Saeb, M., El-Gindi, S.: A DNA-based implementation of YAEA encryption algorithm. In: Proceedings of the second IASTED international conference on computational intelligence 523, 32–36 (2006)
9. Ning, K.: A pseudo DNA cryptography method. (2009), abs/0903.2693
10. Brun, Y.: Nondeterministic polynomial time factoring in the tile assembly model. In: Theoritical computer science, science direct, Elsevier, vol. 395, no. 1, pp. 3–23 (2008)
11. Beaver, D.: Factoring: the DNA solution. In: 4th international conferences on the theory and applications of cryptology. Springer-Verlag, Wollongong, Australia, pp. 419–423 (1994)
12. Li, K., Zou, S., Xv, J.: Fast parallel molecular algorithms for DNA based computation: Solving the elliptic curve discrete logarithm problem over gf(2n). J. Biomed. Biotechnol. Hindawi 2008, 1–10 (2008)
13. Gaurav, G., Nipun, M., Shumpa, C.: DNA computing. The Indian Programmer (2001)
14. Paun, G., Rozenberg, G., Salomaa, A.: DNA computing: new computing paradigms. Springer-Verlag, Berlin (1998)
15. Reif, J.H.: Parallel molecular computations: models and simulations. In: Seventh ACM Symposium on Parallel Algorithms and Architecture, ACM, Santa Barbara 213–223 June-1995 in the US, New Generation Computing 20, no. 3, pp. 217–236
16. Watada, J., Bakar, R.: DNA computing and its applications. In: Eighth international conference on intelligent systems design and applications

17. Cui, G., Liu, Y., Zhang, X.: New direction of data storage: DNA molecular storage technology. Comput. Eng. Appl. **42**(26), 29–32 (2006)
18. Beaver, D.: Factoring: The DNA solution. In: 4th International conferences on the theory and applications of cryptology. Springer-Verlag, Wollongong, Australia, pp. 419–423 (1994)
19. Shannon, C.E.: Communication theory of secrecy systems. Bell Syst. Tech. J. 28-4, pp. 656–715 (1949)
20. Shimanovsky, B., Feng, J., Potkonjak, M.: Hiding data in DNA. Department Computer Science, University of California, Los Angeles
21. Zhang, M., Sabharwal, L., Tao, W.: Interactive DNA sequence and structure design for DNA nanoapplications. IEEE Trans. Nanobiosci. **3**(4), 286–292 (2004)
22. L.H.N.C. for Biomedical Communications: Handbook on Genetic Cells and DNA. USA: National Library of Medicine, National Institutes of Health, Department of Health and Human Services (2010)
23. Taylor, C.T., Risca, V., Bancroft, C.: Hiding messages in DNA microdots. Nature **399**, 533–534 (1999)
24. Lodish, H., Berk, A., Matsudaira, P., Kaiser, C.A., Kreiger, M., Scott, M.P., Lawerance Zipursky, S., Darnell, J.: Molecular cell biology, 5th edn. W.H. Freeman & Company, Chap. 4, pp. 101–145
25. Yunpeng, Z., Zhong, W., Sinnott, R.O.: Index-based symmetric DNA encryption algorithm. In: 2011 4th International congress image signal process, pp. 2290–2294 (2011)
26. Lu, M.X., Lai, X.J., Xiao, G.Z., Qin, L.: Symmetric-key cryptosystem with DNA technology. Sci. China Series F: Inf. Sci. **50**(3), 324–333 (2007)
27. Fujisaki, E., Okamoto, T.: Secure integration of asymmetric and symmetric encryption schemes. NTT Laboratories, LNCS 1666, pp. 537–554, Springer-Verlag, Berlin Heidelberg (1999)
28. Hirabayashi, M., Nishikawa, A., Tanaka, F., Hagiya, M., Kojima, H., Oiwa, K.: Analysis on secure and effective applications of a DNA-based cryptosystem. In: Sixth international conference on bio-inspired computing: theories and applications, pp. 205–210 (2011)
29. Anam, B., Yorkshire, W.: Review on the advancements of DNA cryptography, pp. 25–27 (2010)
30. Cui, G., Cuiling, L., Haobin, L., Xiaoguang, L.: DNA computing and its application to information security field. In: IEEE fifth international conference on natural computation, Tianjian, China, pp. 43–47 (2009)
31. Tornea, O., Borda, M.E.: DNA cryptographic algorithms, international conference on advancements of medicine and health care through technology. In: IFMBE Proceedings 26, pp. 223–226 (2009)
32. Cui, G., Qin, L., Wang, Y., Zhang, X.: An encryption scheme using DNA technology. In: IEEE 3rd international conference on bio- inspired computing: theories and applications (BICTA08), Adelaid, SA, Australia, pp. 37–42 2008
33. Menezes, A., Oorschot, P., Vanstone, S.: Handbook of applied cryptography. CRC Press (1996)
34. Desai, A.: Secure against chosen-ciphertext attack. Department of Computer Science & Engineering, University of California at San Diego, USA (2000)
35. Adleman, L.M.: Molecular computation of solutions to combinatorial problems. Science **266**, 1021–1024 (1994)
36. Head, T.: Splicing schemes and DNA. Lindenmayer systems; impact on theoretical computer science and developmental biology, pp. 371–383 (1992)
37. Pixton, D.: Regularity of splicing languages. Discrete Appl. Math. **69**(1–2), 101–124 (1996)
38. Pixton: Regular splicing systems. Manuscript (1995)
39. Head, T.: Formal language theory and DNA: an analysis of the generative capacity of specific recombinant behaviors. Bull. Math. Biol. **49**(6), 737–759 (1987)

Genetic Algorithm Using Guide Tree in Mutation Operator for Solving Multiple Sequence Alignment

Rohit Kumar Yadav and Haider Banka

Abstract An improved mutation operator in genetic algorithm for solving multiple sequence alignment problems is proposed. In this step, the UPGMA method is used to generate the guide tree where two different matrices such as edit distance or dynamic distance have been used. The performance of the proposed method has been tested on Bali base with some of the existing methods such as, HMMT, DIALIGN, ML–PIMA, and PILEUP8. It has been observed that the proposed method perform better in most of the cases.

Keywords Multiple sequence alignment (MSA) · Genetic algorithm (GA) · Dynamic distance · Edit distance · UPGMA

1 Introduction

MSA is one of the most crucial mechanism for inspecting biological properties. It is very important and challenging task to solve MSA problem in an efficient manner. MSA of amino acid sequence or protein sequence helps to find primary or secondary structure of molecular biology. MSA can also be used to predict functions and molecular structures [1] of same family. Everyone knows that MSA problem can be solved in $O(n^k)$ time complexity, where n is the number of sequences and k is length of sequence using dynamic programming (DP) [2, 3]. Even DP can solve problem in optimal way. But it is useful for only 2 or 3 sequences. In biology,

R.K. Yadav (✉) · H. Banka
Department of Computer Science and Engineering, Indian School of Mines,
Dhanbad 826004, India
e-mail: rohit.ism.123@gmail.com

H. Banka
e-mail: hbanka2002@yahoo.com

© Springer India 2016

R. Chaki et al. (eds.), *Advanced Computing and Systems for Security*,
Advances in Intelligent Systems and Computing 395,
DOI 10.1007/978-81-322-2650-5_10

molecular sequences are very rich in terms of numbers and lengths. Due to this property of molecular sequence, MSA problems lead to NP-hard [4]. So we have to align multiple sequences within limited time complexity. There are many methods have been developed to solve MSA problem in limited resources. Genetic algorithm (GA) is most important tool to solve NP-hard problem. In past, many of them use GA to find good alignments but it is not necessary to give optimal alignment [2].

There are several methods such as classical, progressive, and iterative used to solve MSA problems. Global or local alignment strategies are used in all methods such as classical, progressive, and iterative. These two methods have different properties. In global alignment, sequences are aligned over full length. But in the local alignment, first we have to find the region of similarity then aligned this part of sequences [5]. Both global and local alignments have some advantage and disadvantage. As example local alignment is preferable but sometimes it is more difficult to find the region of similarity. Needleman and Wunsch [6] algorithm is based on global alignment technique which is used dynamic programming (DP). The Smith-Waterman algorithm [7] is based on local alignment technique which is also used dynamic programming (DP). Dynamic programming method [6] gives optimal alignment but it is limited for two or three sequences only. When the number of sequences increases in the alignment process complexity of dynamic programming grows exponentially [8]. Since MSA has more number of sequences to be aligned. Hence MSA is NP-hard problem [9]. The progressive alignment method [10] is iteratively use of Needleman and Wunsch algorithm to find a phylogenetic tree to represent the relationship between two sequences. In the progressive alignment method, we use guide tree for finding initial alignment. The problem behind this method is if we aligned more divergent sequence in initial stage, we cannot improve in later stage. So that difficulty with this method is generally converging to local optima [11]. To overcome such situations researchers switch to iterative or stochastic process [12–14].

In this paper, we introduce a new mechanism in GA, it uses guide tree method in mutation operator. To maximize the matching of protein sequence in MSA, we use well-known PAM [15] score matrix for finding the fitness value of particular solution. To prove the quality of our algorithm, we compare some other well-known existence methods such as DIALI, ML-PIMA, HMMT, and PILEUP-8. For comparison, we use Bali base benchmark dataset. We use PAM matrix for fitness evaluation. For comparing with other well-known existence methods, we calculate corresponding Baliscore of best score.

This paper is organized as follows. In Sect. 2, we discuss about propose method. A concise presentation of test dataset and experimental analysis of propose method is discussed in Sects. 3 and 4, respectively. In the last section, conclusions are provided.

2 Proposed Algorithm

There are various steps in our proposed method. Initial generation, generates new generation using genetic operators and the stopping criteria. All steps of our proposed method are describe below.

2.1 Initial Generation

The aim of this step is to generate initial solutions. In this step, we generate initial solution by putting the gap between residues. Each individual in the population is a candidate solution of MSA problem. Each alignment is represented as two-dimensional matrix of binary strings with each row representing an input sequence and each column representing a position in the alignment. The value of 1 in a bit indicates an inserted gap and the value of 0 indicates a character in the original sequence. For each row in the matrix, we allow 20 % gap. Figure 1 is an example of an individual solution. Simply according to Fig. 1, first is the initial input of MSA problem. After that we put gap randomly in these sequences. This is the initial alignment of given MSA. In Encoding scheme, we simply put the 0 where protein element and put 1 where gap. In chromosome representation, we take binary value of each column from top to bottom. As in our figure 000 is the first column. Binary value of this quantity is 0. Hence, the value of first column is 0. Similarly in fourth column from top to bottom 100 and the binary value of this quantity is 4. Hence, the value of fourth column is 4. According to this method,

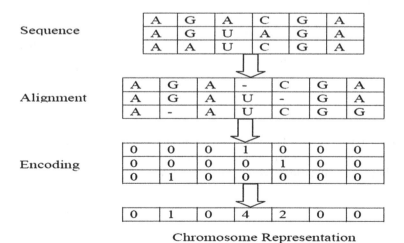

Fig. 1 An example of individuals and chromosome representation of individual solution of MSA

chromosome is 0 1 0 4 2 0 0. Hence, the number of elements in chromosome is equal to the number of column presents in encoding scheme.

2.2 Fitness

The SPM is commonly used as a fitness measure for multiple sequence alignments. Here each column is scored by summing the scores of each pair of symbols. The score of the entire alignment is then summed over all columns score using Eq. 1

$$S = \sum_{l=1}^{L} S_l \, Where \;\; S_l = \sum_{i=1}^{N-1} \sum_{j=i+1}^{N} W_{ij} \text{Cost}\,(A_i, A_j) \qquad (1)$$

Here, S is the cost of multiple alignments. L is the length (columns) of the alignment. S_l is the cost of lth column of L length. N is the number of sequences. W_{ij} is the weight of sequence i and j. It defines diversity between two sequences.

$W_{ij} = $ (Number of Mismatch character in the alignment)$/$(Total align length)

Cost (A_i, A_j) is the alignment score between the two aligned sequences A_i and A_j. When $A_i \neq$ - and $A_j \neq$ - then Cost (A_i, A_j) is determined from the PAM matrix. Also when $A_i =$ - and $A_j =$ - then Cost $(A_i, A_j) = 0$. Finally, when $A_i =$ - and $A_j \neq$ - or $A_i \neq$ - and $A_j =$ - then Cost $(A_i, A_j) = 1$.

2.3 Child Generation

For each individual in the initial population, the SPM score is calculated. To generate a child population of 100 individual solutions in next generation, the following genetic operators are used.

2.3.1 Selection

In this selection, we used tournament selection and the size of this selection is two. First choose two individual solution randomly after that we choose one individual solution according to SPM score of individuals.

2.3.2   Crossover

Two individuals are selected through selection process for crossover. This is implemented [16] in Fig. 2. In this process, first we choose randomly one point in a solution. In Fig. 2 a, "*" is shown in column 3 of parent a. We divide parent "a" from this point. We can also divide parent b into such a way each row of the first piece has the same number of elements as the first piece of the first parent. Now we can interchange of these two pieces of parents to generate child. The whole process of crossover is shown in Fig. 2.

2.3.3   Mutation

In the mutation process, one individual is selected randomly from whole population. After that we find two distance matrix dynamic distance and edit distance.

(A) Dynamic distance—This distance calculates between two sequences. Distance is calculated ratio of number of mismatch elements in these sequences and total length of these aligned sequences.

Dynamic distance = Number of mismatch element/ total align length

We are taking one example of MSA problem to show how we calculate dynamic distance matrix.

ABCD_
_BCDE
CDEF_
_DEFG

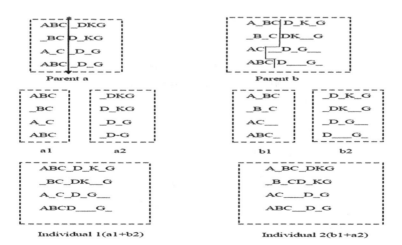

Fig. 2   An example of one point crossover

First, we find pair wise alignment of each pair. According to this alignment, we can find distance between these two sequences. Then first pair (1, 2) is

ABCD_
_BCDE

Since the number of mismatch element (first and last column) in these sequences is 2 and total alignment length is 5. So distance between sequence first pair is 2/5.

Similarly, the second pair (1, 3) is

ABCD_ _
_ _CDEF

Distance between second pair is 4/6.

Similarly, we can find distance between all pairs of this given MSA

Distance between third pair is 6/7.

Distance between fourth pair is 2/5.

Distance between fifth pair is 4/6.

Distance between sixth pair is 2/5. It is shown in Fig. 3.

(B) Edit distance—The minimum Edit distance between two sequences is the minimum number of editing operations (insertion, deletion, and substitution) needed to transform one into other sequences. We are taking same example of MSA problem to show how we calculate edit distance matrix.

ABCD_
_BCDE
CDEF_
_DEFG

First, we find pair wise alignment of each pair. According to this alignment, we can find distance between these two sequences. Then first pair (1, 2) is

ABCD_
_ BCDE
I   D

Here, I stand for insertion and D stand for deletion. Then two minimum edit operations need for transform first sequence to second sequence. Then the minimum edit distance between pair (1, 2) is 2.

	1	2	3
2	2/5		
3	4/6	2/5	
4	6/7	4/6	2/5

Fig. 3 Dynamic distance matrix

Similarly, we can find edit distance between pair (1, 3) is

ABCD_ _
_ _ CDEF
I I   DD

Hence minimum edit distance between pair (1, 3) is 4.
Similarly, we can find distance between all pairs of given MSA.
Distance between third pair (1, 4) is 6.

ABCD_ _ _
_ _ _ DEFG

Distance between fourth pair (2, 3) is 2.

BCDE_
_CDEF

Distance between fifth pair (2, 4) is 4.

BCDE_ _
_ _DEFG

Distance between sixth pair (3, 4) is 2.

CDEF_
_DEFG

It is shown in Fig. 4.
First Cycle:
We choose randomly dynamic distance matrix or edit distance matrix. After that
we use distance matrix, we constructed guide tree with the help of UPGMA [17].
According to the edit distance matrix, we can find guide tree. Since distance
between pairs (1, 2) (2, 3), and (3, 4) is small. So we take any of these pairs. In this
particular, we are taking pair (1, 2)

The Alignment of pair (1, 2) is

ABCD_
_BCDE

	1	2	3
2	2		
3	4	2	
4	6	4	2

Fig. 4 Edit Distance Matrix

Now new distance is

Dist (1, 2), 3 = (dist13 + dist23) / 2 = 3

Dist (1, 2), 4 = (dist14 + dist24) / 2 = 5

Second Cycle:

	1, 2	3
3	3	
4	5	2

3
4

The Alignment of pair (3, 4) is

CDEF_
_DEFG

Third Cycle:

	1, 2
3, 4	4

Now combined all sequences

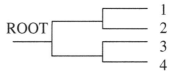

ROOT
1
2
3
4

Then complete alignment is

ABCD_
_BCDE
  CDEF_
  _DEFG

For a valid solution, we put gap in all vacant places. Then complete alignment is

ABCD_ _ _
BCDE _
_ _CDEF_
_ _ _DEFG

The Flow chart of this mutation process is given in Fig. 5.

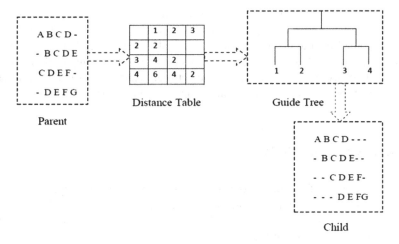

Fig. 5 An example of mutation process

2.4 Termination Condition

The best solution in each generation is recorded. If the best solution remains unchanged in next 50 consecutive generations, then stop this algorithm. This observation is based on experimental review. We tested our algorithm up to 200 generations after getting best solution. We saw that the best solution hardly change and the variation of average solution per generation also very low.

3 Test Dataset

We have taken a huge number of datasets from Bali base benchmark datasets for comparison between propose and other existing methods. There are two versions of Bali base benchmark datasets. Bali base version 1.0 [18] contains 142 different types of datasets. Bali base version 1.0 contains 1000 sequences for 142 datasets. Bali base version 2.0 [19] is an extended version of Bali base version 1.0. In the Bali base version 2.0 contains 167 different varieties of datasets over 2100 sequences. There are eight different types of reference set in Bali base version 2.0. Small number of innermost sequences is contained in the reference set 1. Orphan or distinct sequences are containing in reference set 2. A pair of dissimilar subfamily are contain in reference set 3. Long-terminal extensions and large internal insertions and deletions are containing in reference set 4 and reference set 5, respectively. Last, references 6–8 contain test case problems where the sequences are repeated and the domains are inverted. Bali score is a score which lies between 0 and 1. Bali score is an open source program which measures the accuracy of our alignment. It is available on

BALIBASE VERSION 2.0. It compares between two alignments. First one is Bali base Dataset which is manually aligned and second one is alignment of this dataset which is come from our method. If the manually alignment file and our output file is same, then score is 1. If the manually alignment file and our output file is totally different, then score is 0. It gives the value between 0 and 1 according to similarity between Bali base manually alignment file and our output file.

4 Experimental Study

We examine the performance of our propose method. Then we compare our result with other well-known existence methods. In this proposed method, first runs the algorithm ten times and takes best results out of these ten runs. After that we find corresponding Bali score of the best result.

4.1 Effect of Operators and Fitness Function

The proposed approach uses a modified mutation operator and fitness function. This is helpful for getting good results. We design two set of experiments to examine the reaction of constituent. Proposed method runs with improve mutation operator and fitness function in the first set and in the second set we run our proposed approach with simple mutation and fitness function. We have taken randomly 5 datasets for these experiments 3 from reference set 2 and 2 from reference set 3. Each dataset run with proposed approach with two different configurations. First one is proposed with improve mutation operator and second one is proposed with simple mutation operator. Each datasets run ten times with this proposed algorithm. Corresponding Bali score of the best SPM solution found, proposed method with improved mutation operator and improved fitness function performs better than proposed method with simple operator for all datasets. These experiments prove the dominance of our proposed fitness function and the improved mutation operator. The details of these experiments are reported in Table 1.

Table 1 Performance test GA with simple and improved mutation operator	Datasets	Proposed with improved operator	Proposed with simple operator
	1aboA	**0.399**	0.208
	1csy	**0.509**	0.272
	1wit	**0.288**	0.235
	1idy	**0.423**	0.250
	1uky	**0.207**	0.159
	Avg score	**0.365**	0.224

4.2 Comparing Proposed Method with Existence Methods

We have considered 15 datasets from reference 2 and 5 datasets from reference set 3 [20]. We take the results of other well-known existence methods from RBT-GA which is presented in paper [20]. Tables 2 and 3 show that experimental results of reference set 2 and reference set 3, respectively.

Performance of Proposed method in reference 2: We have taken 15 datasets from reference set 2 which are significantly different in terms of length and number of sequence. Proposed method performs better with different type of datasets. We compare proposed method with well-known existence methods such as DIALIGN, HMMT, PILEUP-8, and ML-PIMA with respect to Bali score to determine the performance of proposed method. Proposed method has found better 9 test cases and 6 test cases where proposed method not perform well close to the best solution.

Table 2 Experimental results on reference 2 datasets of Bali Base 2.0

Name of datasets	DIALI	HMMT	ML-PIMA	PILEUP-8	Proposed
1aboA	0.384	**0.724**	0.22	0.000	0.399
1idy	0.000	0.353	0.000	0.000	**0.559**
1csy	0.000	0.000	0.000	0.114	**0.509**
1r69	0.675	0.000	**0.675**	0.45	0.251
1tvxA	0.000	0.276	0.241	0.345	**0.534**
1tgxA	0.63	0.622	0.543	0.318	**0.705**
1ubi	0.000	0.053	0.129	0.000	**0.624**
1wit	**0.724**	0.641	0.463	0.476	0.288
2trx	0.734	0.739	0.702	**0.87**	0.787
1sbp	0.043	0.214	0.054	0.177	**0.221**
1havA	0.000	0.194	0.238	0.493	**0.495**
1uky	0.216	0.395	0.306	**0.562**	0.243
2pia	0.612	0.647	0.695	**0.766**	0.179
3grs	0.350	0.141	0.211	0.159	**0.395**
kinase	0.692	0.749	0.651	0.799	**0.898**
Avg score	0.337	0.383	0.351	0.368	**0.458**

Table 3 Experimental results on reference 3 datasets of Bali Base 2.0

Name of datasets	DIALI	HMMT	ML-PIMA	PILEUP-8	PROPOSED
1idy	0.000	0.227	0.000	0.000	**0.423**
1r69	0.524	0.000	**0.905**	0.000	0.425
1ubi	0.000	0.366	0.000	0.268	**0.598**
1wit	**0.500**	0.323	0.323	0.210	0.290
1uky	0.139	0.037	0.148	0.083	**0.207**
Avg score	0.232	0.190	0.275	0.112	**0.388**

Table 2 shows that overall performance of proposed method achieves better than other existence methods. Finally, we can say that proposed method perform better than all other methods.

Performance of Proposed method in reference 3: Reference 3 contains dissimilar type of datasets where the residue identities between sequences are less than 25 %. We consider only 5 dataset out of 12. Table 3 shows that in 3 test cases out of 5 test cases proposed method perform better than other existence methods. Although the proposed method did not achieve high-accuracy solutions in two test cases, the average performance of this method was clearly better than the others as shown in Table 3.

5 Conclusion

In this paper, we use an improved mutation operator in GA for solving MSA problem. This mutation operator has been generated using of guide tree. A significant number of experiments have been done to prove the quality of our proposed method. The proposed method was run with simple mutation operator as define in simple GA to prove the superiority of our proposed mutation operator and improved fitness function. The proposed method was optimized based on the sum of pair score, and this is used as a fitness function. A number of benchmark datasets from Balibse 2.0 is used in this paper. Therefore, the corresponding Bali score of best score out of ten runs was used to compare with other methods. The experimental results show that proposed method performs better and/or competitive most of time. Although the solution of proposed method was not always the best for some test cases. However, it is always close to the best. The improved mutation operator is reason behind the proposed method is better than other existence methods. Hence after the experimental analysis, we can say that the proposed method is better than other existence method.

Acknowledgments The work was partially supported by CSIR grant no. 22(0586)/12/EMR-11.

References

1. Gusfield, D.: Algorithms on Strings, Trees, and Sequences. Cambridge University Press, New York, Computer Science and Computational Biology (1997)
2. Feng, D.F., Johnson, M.S., Doolittle, R.F.: Aligning amino acid sequences: comparison of commonly used methods. J. Mol. Evol. **21**(2), 112–125 (1985)
3. Holland, J.H.: Adaptation in natural and artificial systems. University of Michigan Press, Ann Arbor (1975)
4. Michalewicz, Z., Genetic algorithms + data structures = evolution programs—Third, Revised and Extended Edition, 3 edn. Springer, Berlin (1996)

5. Thompson, J.D., Plewniak, F., Poch, O.: A comprehensive comparison of multiple sequence alignment programs. Nucleic Acids Res. **27**(13), 2682–2690 (1999)
6. Needleman, S.B., Wunsch, C.D.: A general method applicable to the search for similarities in the amino acid sequence of two proteins. J. Mol. Biol. **48**(3), 443–453 (1970)
7. Smith, T.F., Waterman, M.S.: Identification of common molecular subsequences. J. Mol. Biol. **147**(1), 195–197 (1981)
8. Stoye, J., Perrey, S.W., Dress, A.W.M.: Improving the divide-and conquer approach to sum-of-pairs multiple sequence alignment. Appl. Math. Lett. **10**(2), 67–73 (1997)
9. Bonizzoni, P., Vedova, G.D.: The complexity of multiple sequence alignment with SP-score that is a metric. Theor. Comp. Sci. **259**, 63–79 (2001)
10. Feng, D.F., Dolittle, R.F.: Progressive sequence alignment as a prerequisite to correct phylogenetic trees. J. Mol. Evol. **25**(4), 351–360 (1987)
11. Thompson, J.D., Higgins, D.G., Gibson, T.J.: CLUSTALW: Improving the sensitivity of progressive multiple sequence alignment through sequence weighting, position-specific gap penalties and weight matrix choice. Nucleic Acids Res. **22**(22), 4673–4680 (1994)
12. Gotoh, O.: Optimal alignment between groups of sequences and its application to multiple sequence alignment. Comput. Appl. Biosci. **9**(3), 361–370 (1993)
13. Lukashin, A.V., Engelbrecht, J., Brunak, S.: Multiple alignment using simulated annealing: branch point definition in human mRNA splicing. Nucleic Acids Res. **20**(10), 2511–2516 (1992)
14. Lawrence, C.E., Altschul, S.F., Boguski, M.S., Liu, J.S., Neuwald, A.F., Wooton, J.C.: Detecting subtle sequence signals: q Gibbs sampling strategy for multiple alignment. Science **262**, 208–214 (1993)
15. Dayhoff, M.O., Schwartz, R.M., Orcutt, B.C., A model of evolutionary change in proteins, Atlas Protein Sequence Structure, vol. 5, no. 3, pp. 345–351 (1978)
16. Shyu, C., Sheneman, L., Foster, J.A.: Multiple sequence alignment with evolutionary computation. Genet. Program. Evol. Mech. **5**, 121–144 (2004)
17. Sneath, P.H.A., Sokal, R.R.: Taxonomic structure. In: Taxonomy, Numerical (ed.) San Francisco, pp. 188–308. Freeman, CA (1973)
18. Thompson, J.D., Plewniak, F., Poch, O., Bali, B.A.S.E.: A benchmark alignments database for the evaluation of multiple sequence alignment programs. Bioinformatics **15**(1), 87–88 (1999)
19. Bahr, A., Thompson, J.D., Thierry, J.C., Poch, O.: BALIBASE (benchmark alignment database): enhancements for repeats, trans membrane sequences and circular permutation. Nucleic Acids Res. **29**(1), 323–326 (2000)
20. Taheri, J., and Zomaya, A.Y., RBT-GA: A novel metaheuristic for solving the multiple sequence alignment problem. BMC Genomics **10**(1) 1–11 (2009)

A Comparative Analysis of Image Segmentation Techniques Toward Automatic Risk Prediction of Solitary Pulmonary Nodules

Jhilam Mukherjee, Soharab Hossain Shaikh, Madhuchanda Kar and Amlan Chakrabarti

Abstract Lung cancer is considered as a leading cause of death throughout the globe. Manual interpretation of cancer detection is time consuming and thus increases the death rate. With the help of improvement in medical imaging technology, a computer-aided diagnostics system could be an aid to combat this disease. Automatic segmentation of a region of interest is one of the most challenging problem in medical image analysis. An inaccurate segmentation of solitary pulmonary nodule may lead to an erroneous prediction of the disease. In this paper, we perform a comparative study among the available segmentation techniques, which can automatically segment the solitary pulmonary nodules from high-resolution computed tomography (CT) images and then we propose a computerized lung nodule risk prediction model based on the best segmentation technique.

J. Mukherjee (✉) · A. Chakrabarti
A.K. Choudhury School of Information Technology, University of Calcutta,
Kolkata, India
e-mail: jhilam.mukherjee20@gmail.com

A. Chakrabarti
e-mail: acakcs@caluniv.ac.in

S.H. Shaikh
Department of Computer Science and Engineering, NIIT University,
Neemrana, India
e-mail: Soharab.Hossain@niituniversity.in

M. Kar
Department of Oncology, Peerless Hospital and Research Centre Ltd.,
Kolkata, India
e-mail: madhuchandakar@yahoo.com

© Springer India 2016
R. Chaki et al. (eds.), *Advanced Computing and Systems for Security*,
Advances in Intelligent Systems and Computing 395,
DOI 10.1007/978-81-322-2650-5_11

Fig. 1 Solitary pulmonary
nodule [8]

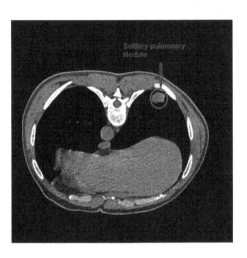

1   Introduction

Lung cancer is basically of two types, primary lung cancer and metastasis lung cancer. Primary lung cancer reveals itself as solitary pulmonary nodule (SPN), which can be found in CT scans. They are little bright spots enclosed by lung parenchyma, with gray-values very similar to those of blood vessels in the lungs and normally spherical in shape as illustrated in Fig. 1. Those that are bound to the pleural surface are known as juxtapleural nodule, which are roughly hemispherical and responsible for metastasis lung cancer. These wounds can be discovered on CT images even if they are only a few millimeters in diameter. Although the immense majority of pulmonary nodules is nonmalignant and do not require treatment, detection of such nodules is the first significant step in predicting risk in lung nodules at an early point.

The sensing and diagnosis of suspicious lung nodules in CT images is one of the main challenges in medical image analysis. In recent days, early detection and diagnosis of solitary pulmonary nodule has received massive attention. With the improvement of CT imaging techniques, nowadays a thin-section CT scan can generate almost 400 images with 1-mm thickness. The radiologists spend a huge time to decide about the appropriate image slice required for diagnosis and that could lead to a time consuming process and thus increases variability in measurement. A computer-aided diagnostic procedure can be involved to bring down the workload of a radiologist and to increase the precision of detection with high sensitivity and low false positive rate.

The rest of the paper is organized as follows, in Sect. 2 we discuss the relevant existing research works. Section 3 points out the detailed methodology used in this study. Section 4 reveals the outcome of this study. Section 5 concludes the paper with a brief on the future scope of this study.

2 Related Work

It has been seen that lung nodules have greater intensity value than lung parenchyma. This difference in intensity value can easily separate it from its surrounding structure. Various researchers have proposed various nodule detection methodologies using this intensity difference. A successful attempt of intensity-based approach is designed by Sousa et al. [1]. This scheme is implemented by simulating seven successive steps of image processing techniques. Region grow algorithm is used for artifacts removal and to extract the lung region. A rolling ball algorithm is used to reconstruct the lung wall. Nodules are classified using support vector machine (SVM) followed by a bifurcation procedure to remove the blood vessels. Lung nodule detection using K-means clustering technique has been proposed by Gurcan et al. [2]. After successful clustering of lung nodules, they have used rule base classifier to classify lung nodules followed by a false positive reduction procedure using linear discriminating analysis.

An autonomous lung nodule detection is proposed by [3] using 3D template matching. Spiral CT images are prepared for template matching after image thresholding and edge enhancement process. Authors have created eight templates of diameter from 3 to 20 mm and trained them using different lung nodule features. 3D template matching procedure is executed using 3D-sum of squared differences (SSD) algorithm and the results are marked as the nodule candidates. Sensitivity of this system is 85.71 % and FPR is 5.4 %. A three phase lung nodule detection methodology is presented in [4]. In first phase, authors have extracted ROI using maximal and minimal density thresholding concept using Hournsfield Unit value which is X-ray attenuation constant used in CT scan procedure. In nodule detection phase, 3D template matching using convolution-based filtering technique. Finally, false positive rate elimination procedure is implemented using a labeling algorithm and the density threshold of surrounding structure gives 0.46/slice. This algorithm is capable to achieve 100 % sensitivity.

A proposal for autonomous lung nodule detection is formulated by Netto et al. [5] using growing natural gas (GNG). At first, the authors selected four seed points and region growing algorithm is applied to extract chest and lung and followed by a lung reconstruction using rolling ball algorithm. In the next step, lung nodules are extracted using GNG algorithm. Lastly, detected structures are classified into nodules and non-nodules using SVM. Average sensitivity reaches to 86 % and FPR is 0.138/scan, but the number of test cases restricted to 39 patients. Another novel lung nodule detection scheme has been recommended by [6]. In this method, nodules are detected using iris filter and LDA to reduce false positive rate.

2.1 Contributions

In this paper, we represent an evaluation of state-of-the-art image segmentation techniques to find out the best option for the prediction of risk of a lung nodule from high-resolution CT images. We have used risk prediction model as stated in [7] for this purpose. Our contributions in this paper are as follows:

- A comparative analysis among different image segmentation techniques and identification of the best technique for lung nodule segmentation from lung CT images.
- Risk prediction of the segmented nodules using multiclass support vector machine classifier.

3 Methodology

At first, a comparative study is performed over collected CT images from hospital [8] for the various segmentation techniques. Risk prediction of lung nodules is performed using multiclass SVM based on the extracted features from the segmented images obtained through the best segmentation technique as described in Fig. 2.

3.1 Image Segmentation

Image Segmentation using Thresholding Technique There exists some intensity difference between lung parenchyma and SPN, hence an appropriate image thresholding technique can easily separate the SPN from its surrounding structures. Otsu global thresholding technique [9] is applied on the input images to create a binary image, which separates the background and lung parenchyma from fat and muscles. Black region of the binarized images indicates intensity value 0 and white region as 1. The binary image may contain air, vessels other than the lung nodules. A morphological operator is used to remove all vessels and air surrounding the

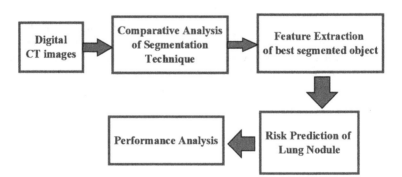

Fig. 2 Work flow of the lung nodule risk quantification system

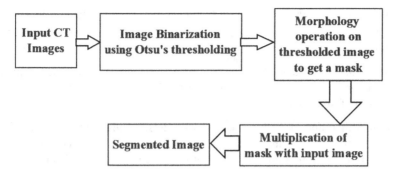

Fig. 3 Segmentation using thresholding techniques

lung. As a result, a hollow, free binary mask is obtained as output. Lastly, the segmented image is obtained by multiplying the original image with the binary mask. Figure 3 and Algorithm 1 describe how binarization and thresholding techniques are used to segment lung nodule from images.

Image Segmentation Using Fast Global Minimization Active Contour
Although, active contour model or snake is one of the most successful model used in image segmentation but one of the drawback of this model is the existence of local minima in active contour energy, which makes initial guess to get satisfactory results. Bresson et al. [10] combines working principle of Rudin–Osher–Fatemi denoising model [11] and Mumford-shah's segmentation model [12], with Geodesic active contour model [13] to overcome this problem.

Algorithm 1 Thresholding and Binarization

Input: HRCT image of Human Lung in DICOM format
Output: Segmented lung image containing only SPN.
  1: Image threshold value is selected using Otsu's thresholding.
  2: Selected threshold value is applied on the input image to obtain a binary image.
  3: Mathematical morphology is applied on the binary image to obtain a hallow free mask.
  4: Nodule segmented images are obtained by multiplying mask with the input image.

Caselles et al. [13] modifies some pitfalls of Kass's original Snake model [14] and proposed a new methodology called geodesic active contour (GAC). The energy minimization of Geodesic active contour is defined as:

$$\min_{C}\left\{ E_{\text{GAC}}(C) = \int_{0}^{L(C)} g(|\nabla I_0(C(s))|)\,\mathrm{d}s \right\} \tag{1}$$

where ds is the Euclidean element of length and $L(C)$ is the length of the curve C and $L(C) = \int_0^{L(C)} ds$. g is an edge indicator function that vanishes at object boundaries such that $g(|\nabla I_0|) = \frac{1}{\beta|\nabla I_0|^2}$, where I_0 is the original image, β is an arbitrary positive constant. This provides the calculus variation of Euler–Lagrange equation of energy functional E_{GAC} and gradient descent method gives the flow that minimizes as fast as possible.

Global minimization of Active contour using ROF model In this algorithm, the authors combine the principle of snake algorithm with Rudin, Osher, and Fatemi's model. The minimization problem associated with ROF model is defined as:

$$\min_u \left\{ E_{\text{ROF}}(u, \lambda) = \int_\Omega |\nabla u| dx + \lambda \int_\Omega (u - f)^2 dx \right\} \tag{2}$$

where f is given image, $\Omega \subset R^N$ is a set representing image domain, $\text{TV}(u)$ is the total variation norm of the function u, and $\lambda > 0$ is an arbitrary parameter. A convex energy was defined for any observed image $f \in L^1(\Omega)$ and any positive parameter λ as

$$E_1(u, \lambda) = \int_\Omega g(x)|\nabla u| dx + \lambda \int_\Omega |u - f| dx \tag{3}$$

The difference between the energy (7) and ROF model are known as weighted TV-norm, $\text{TV}_g(u)$ with a weight function $g(x)$ and a replacement of L^2-norm by L^1-norm. Introduction of a weight function g, in the TV-norm gives the link between the snake of Geometric active contour model and the proposed functional model, as the energy defined in geometric active contour is equal to the TV-norm. g is an edge indicator function, u is a characteristics function. 1_{Ω_C} of a closed set $\Omega_C \subset \Omega$ in which C denotes the boundaries of Ω_C

$$\text{TV}_g(u = 1_{\Omega_C}) = \int_\Omega g(x)|\nabla 1_{\Omega_C}| dx \int_C g(x) ds = E_{\text{GAC}}(C). \tag{4}$$

Global Minimization of Active Contour Model Using Mumford Shah Model
Modified segmentation model described in previous subsection is applicable for binary images. To overcome this constraints, the energy function of geodesic/geometric active contour model is fused with global minimization principal of active contour detection without edge (ACWE) model [15].

For an input image f, Energy minimization of ACWE can be described as follows:

$$\min_{\Omega_{C,c1,c2}} \left\{ E_{\text{ACWE}}(\Omega_{C,c1,c2}, \lambda) = \text{Per}(\Omega_C). \right.$$

$$+ \lambda \int_{\Omega_C} (c1 - f(x))^2 dx$$

$$\left. + \lambda \int_{\Omega\Omega_C} (c1 - f(x))^2 dx \right\}$$

(5)

where Ω_C is a closed subset of image domain Ω Per(Ω_C) is the perimeter of set Ω, λ is an arbitrary constant that controls the trade off between the regularization process and the fidelity of solution with respect to the original image f and $c1, c2 \in R$.

As the minimization problem described in equation is nonconvex and leads to local minima, then the energy of ACWE can be written according to level set function ϕ as:

$$E_{\text{ACWE}}^2(\Phi, c1, c2, \lambda) = \int_\Omega |\nabla H_\epsilon(\Phi)| + \lambda \int_\Omega (H_\epsilon(\Phi)(c1 - f(x))^2 + H_{-\epsilon}(\Phi)(c2 - f(x))^2) dx$$

(6)

Energy minimization for any input image $f \in L^1(\Omega)$ and for any arbitrary constant $\lambda > 0$ can be expressed as:

$$E_2(u, c1, c2, \lambda) := \text{TV}_g(u) + \lambda \int_\Omega r_1(x, c1, c2)\phi dx$$

(7)

Above equation represents the global minimization of energy required for the segmentation. Where u and g are the characteristics and edge indicator function. Figure 4 describes how fast global minimization of active contour algorithm is used to segment lung nodule from images.

Region Grow In region grow algorithm [16], objects are segmented based on a selected points known as seed points. These seed points form a region using some similarity properties like intensity values of neighboring pixels. As dissimilarity appears, growth is ended and a segmented region is obtained. The seed points check the similarity properties along with 8-connected or 4-connected properties of that neighboring pixels. The region is grown iteratively using a threshold value unless there is a mismatch when it stops growing and gives a distinct region. Higher intensity threshold, yields higher region.

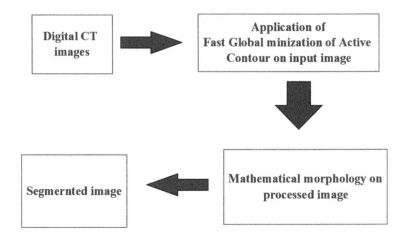

Fig. 4 Segmentation using fast global minimization of active contour

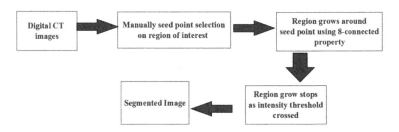

Fig. 5 Segmentation using region grow

Figure 5 and Algorithm 2 describes how region grow algorithm is used to segment lung nodule from images.

Algorithm 2 Region Grow

Input: HRCT image of Human Lung in DICOM format
Output: Segmented lung image containing only SPN.

  1: Select seed points on region
  2: Region grows according to the intensity threshold of surrounding seed points using 8-connected properties of the pixels.
  3: Mean region is calculated.
  4: A segmented algorithm is obtained.

Watershed Transform In the watershed transform [16], the edge information separates the region of interest from its background. In this algorithm, gradient length represents height. Each local minima considered as sink. Watersheds are the boundaries, which separate regions to drain into sink. Let us consider

$N_1, N_2, \ldots, N_R$ are set of points of local minima of an image $I(x, y)$. The surface will be flooded using an integer increment, which varies from $n = \min + 1$ to $n = \max - 1$. Set of coordinate points of catchment basin is represented by $C(N_i)$ and $T[n]$ is the points (s, t) for which $I(s, t) < n$.

$$T[n] = \{(s, t) | I(s, t) < n\} \tag{8}$$

At any step of flood process, algorithm calculates number of points below the flood path. All the coordinating point below the flood path marked as 0 and vice versa.

$C_n(N_i)$ is a set of point of catchment basin are at flooded stage n represents a binary image as

$$C_n(N_i) = C(N_i) \cap T[n] \tag{9}$$

Union of flooded catchment basins at stage n are represented as,

$$C[n] = \bigcup_{i=1}^{R} C_n(N_i) \tag{10}$$

Let $C[\max + 1]$ is the union of all catchment basins and,

$$C[\max + 1] = \bigcup_{i=1}^{R} (N_i) \tag{11}$$

Number of elements of sets $C_n(N_i)$ and $T[n]$ either increases or remains same as the flooding stage increases. Then $C[n - 1]$ can be represented as a subset of $C[n]$ and formed iteratively if the following conditions are satisfied,

1. $q \cap C[n - 1]$ is empty
2. $q \cap C[n - 1]$ contains one connected components of $C[n - 1]$
3. $q \cap C[n - 1]$ contains more than one connected components of $C[n - 1]$

where $q \in Q[n]$ and Q is a set of connected components of $T[n]$.

Condition 1 holds if a new minimum occurs. Condition 2 is possible if q lies on catchment basins. Condition 3 occurs when a crisp separating more than one catchments basin. Figure 6 describes how watershed is used to segment lung nodule from images.

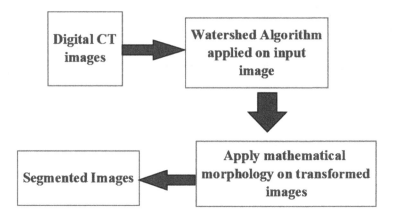

Fig. 6 Segmentation using watershed transform

3.2 Risk Prediction Using Multiclass SVM

According to specialist oncologist, the risk of a lung nodule is divided into three categories namely, benign, malignant, and suspicious. An appropriate pattern classification algorithm can classify the risk of a lung nodule into these three classes. Risk of a lung nodule depends on several features viz., shape, size, calcification, and growth rate. Shape of a lung nodule depends on circularity, moments, and aspect ratio. In this study, we have considered only size and shape to predict the risk of a lung nodule. These feature values are considered as a training set (Figs. 7 and 8).

Standard SVM is designed for a two-class classifier problem. It is a supervised learning tool that classifies the dataset using machine learning theory. In training phase, it takes a data matrix as a input and marks each entity into any one of the two classes. But it cannot classify a data having classes more than two.

Fig. 7 Visual verification result in coronal [8], **a** input image, **b** ground truth image, **c** thresholding, **d** ACM, **e** region grow, **f** watershed

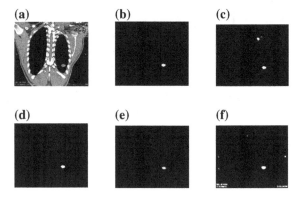

Fig. 8 Visual verification result in sagittal view [8], **a** input image, **b** ground truth image, **c** thresholding, **d** ACM, **e** region grow, **f** watershed

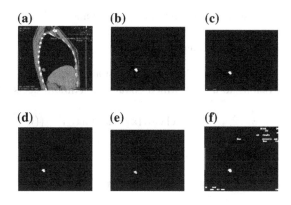

Our problem is a three-class problem and it cannot be handled efficiently by SVM. This problem can be handled using multiclass support vector machine. One against all approach is the most used techniques to contrast multiclass SVM [17]. In this method, we need to construct k-SVM Model where k denotes number of class, $k = 3$ for our case. The ith SVM are trained with positive label and others with negative label. Each data are trained considering malignant as positive level and suspicious and benign as negative level. Then for l training data $(x_1, y_1), \ldots, (x_l, y_l)$ where $x_i \in R^n$ and $y_i \in \{i, \ldots, k\}$ is the class of x_i. Hence ith SVM solves following problem.

$$\min_{w^i, b^i, \xi^i} \frac{1}{2}(w^i)^T w^i + C \sum_{j=1}^{l} \xi^i$$

$$(w^i)^T \phi(x_j) + b^i \geq 1 - \xi_j^i, \quad \text{if } y_j = i$$

$$(w^i)^T \phi(x_j) + b^i \leq -1 + \xi_j^i, \quad \text{if } y_j \neq i$$

$$\xi_j^i \geq 0, j = 1, \ldots, l$$

where C is the regulatory parameter that controls the misclassification error of the training data, ξ is slack variable that reduces the constraints of inequalities of a nonseparable case. Training data x_i are mapped to a higher dimension subspace by function ϕ. Minimization of $\frac{1}{2}(w^i)^T w^i$ means maximizing the margins of two group of data. If the data are not linearly separable, then a penalty term $C \sum_{j=1}^{l} \xi^i$ is used to reduce number of training errors. x_i is the pattern among which the problem is classified. Here $i = 3$ and the value of x_i are malignant, benign, and suspicious. y_i is the feature vectors or training data. We have used 20 nodule feature data as a training set and each feature vector contain five individual features value namely size, aspect ratio, circularity, and moments. On the other hand, rest of 20 nodule features are tested for classification. In both cases, the dimensionality of feature vector and training set are same.

Solving Eq. 3.2 we get k decisions functions as, $(w^1)^T \phi(x) + b^1 \vdots (w^k)^T \phi(x) + b^k$ x belong to a class which contains maximum decision functions.

$$\text{classo fx} = \text{argmax}_{i=1,\dots,k}((w^i)^T \phi(x) + b^i) \tag{12}$$

4 Results and Discussions

4.1 Experimental Dataset

The images used in this study were provided by Peerless Hospitex Hospital, Kolkata [8]. These 25 patient images were obtained from 25 different real cancer patients consisting of 40 nodules. The images were acquired through a 16-row multidetector CT scanner under the following conditions: a single-breath hold with tube voltage 120 kVp, tube current 100 mA, and 0.5 mm collimated during suspended full inspiration. Average number of slices is 415. Image matrix size 512 × 512 in DICOM (Digital Image COmmunication in Medicine) format. In this research, we have selected coronal and sagittal view of CT data.

4.2 Ground Truth Image Creation

We have created ground truth images using majority voting algorithm, as described in [18] for data validation purposes. We have considered all images of the three views of a CT images. These ground truth images are throughly verified by a specialist oncologist and a radiologist who are actively engaged in this project.

4.3 Comparative Study of Image Segmentation

In this section, we represent the result of different image segmentation techniques viz., thresholding, fast global minimization of active contour (ACM), region Grow, and watershed transform (WST). Some standard metrics like misclassification error

(ME), relative area error (RAE) [19], and correlation coefficient (CoC) [20] have been carried out to select best image segmentation technique for risk prediction of lung nodule from CT images.

- Misclassification Error: Misclassification Error represents percentage of background pixels incorrectly allocate to foreground and vice versa, can be expressed as

$$\text{ME} = 1 - \frac{|B_O \cap B_T| + |F_O \cap F_T|}{|B_O \cap F_O|} \qquad (13)$$

where B_O, F_O are the background and foreground of the ground truth image and B_T, F_T are the background and foreground of the test image. $|.|$ is the cardinality of the set. ME value 0 denotes perfectly segmented image and 1 represents totally different image.

- Relative Foreground Area Error: This measurement is quantified using area feature of test image and ground-truth image.

$$\text{RAE} = \begin{cases} \frac{A_O - A_T}{A_O} & \text{if } A_T < A_O \\ \frac{A_T - A_O}{A_O} & \text{if } A_T \geq A_O \end{cases} \qquad (14)$$

where A_O and A_T are the area of ground-truth image and test image. Then, for a perfectly segmented region RAE is 0 and vice versa.

- Correlation Coefficient: It is a measurement of identity between two images, value lies between 0 and 1.1 represents an identical image and vice versa. Mathematically, CoC can be defined as,

$$\text{CoC} = \frac{\sum_{i=1}^{x} (A(i) - \overline{A}(i))(B(i) - \overline{B}(i))}{\sqrt{\sum_{i=1}^{x} (A(i) - \overline{A}(i))(B(i) - \overline{B}(i))}} \qquad (15)$$

where $A(i)$ and $B(i)$ are ground truth image and segmented images and $\overline{A}(i)$ and $\overline{B}(i)$ are respected mean of this images.

Confidence Interval Confidence level is a statistical measurement that describes the range of a sample parameter with a predefined confidence. Confidence interval of a sample can be calculated as,

Algorithm 3 Confidence Interval

Input: sample Dataset

Output: Range of Sample

1: calculate sample size n

2: calculate sample mean & Standard Deviation

3: Measure standard Error

$$standard\ error = standard\ deviation/\sqrt{(n)} \tag{16}$$

4: Select confidence level to assign critical value

5: Calculate α as,

$$\alpha = 1 - \frac{confidencelevel}{100} \tag{17}$$

6: Calculate critical probability as,

$$critical\ probability = 1 - \alpha/2; \tag{18}$$

7: Calculate degree of freedom as

$$degreeoffreedom = n - 1; \tag{19}$$

8: Calculate Margin of Error

$$margin\ error = standard\ error * critical\ value \tag{20}$$

9: Calculate Confidence Interval as

$$Confidence\ interval = samplemean \pm margin\ error \tag{21}$$

As, the intensity value of a lung nodule is very similar to blood vessels, then some times an image segmentation algorithm also segment blood vessels along with lung nodule and erroneously classified it into malignancy and benignity. To overcome this problem, researchers have to select an appropriate image segmentation algorithm that can segment only lung nodule and not other irrelevant objects of lung parenchyma. Table 1 shows that binarization and thresholding techniques and region grow algorithm segments object is similar to number of object present in ground truth image and watershed algorithm behaves worst. In Fig. 9, the plot of image binarization and thresholding technique and region grow algorithm coincides with plot of ground truth image, which clearly indicates that these two algorithms in case of object detection, but in region grow algorithm, a seed point is selected manually. This means this algorithm is in contradiction to the concept of automation.

According to the definition of ME, a good image segmentation algorithm yields less ME value. Table 2 and Fig. 10 depicts that thresholding and binarization gives less ME value, then thresholding, region grow, and watershed algorithm. This measurement also concludes the effectiveness of thresholding and binarization over other segmentation techniques toward automatic segmentation of lung nodule from CT images.

Table 1 Object detection by each techniques

Image name	Ground truth	ACM	Thresholding	Watershed	Region grow
Image1	1	2	1	5	1
Image2	3	5	3	8	3
Image3	1	2	1	4	1
Image4	3	4	3	9	3
Image5	2	3	2	5	2
Image6	1	1	1	1	1
Image7	1	1	1	1	1
Image8	2	2	2	4	2

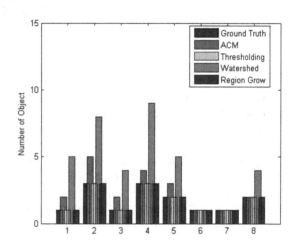

Fig. 9 Object detection by segmentation techniques

Table 2 Misclassification error

Image name	ACM	Thresholding	Watershed	Region grow
Image1	0.0700	0.0651	0.0584	0.0618
Image2	0.0387	0.0356	0.0444	0.0544
Image3	0.1851	0.1687	0.1806	0.1749
Image4	0.1246	0.1026	0.1244	0.1205
Image5	0.0873	0.0845	0.1007	0.1009
Image6	0.1603	0.1304	0.1314	0.1326
Image7	0.2290	0.2009	0.2406	0.2044
Image8	0.1733	0.1671	0.1955	0.1713

An output of a good segmentation algorithm is very much identical with its ground truth image and gives correlation coefficient value very near to one. According to Table 4 image binarization and thresholding techniques gives CoC value very near to one and indicates best segmentation algorithm among four.

Fig. 10 Misclassification
error

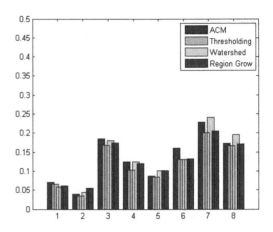

In this study, we have considered Confidence Interval as, 95 % and we have applied the above algorithm to get confidence interval for misclassification error between 0.1085 and 0.1490, relative area error between 0.1085 and 0.1490, correlation coefficient between 0.9984 and 0.9990.

Figures 7 and 8 shows segmentation results of different techniques. From this diagram, it is clear that region grow algorithm segments less number of nodule candidates and this may mislead to erroneous prediction of the disease. Table 1 and Fig. 9 shows that number of object detection by image binarization and thresholding techniques gives nearly same as given in case of ground truth images, which clearly indicates the efficiency of this algorithm. Misclassification error and relative area error define that the value of these two metrics will be equal to zero in a perfectly identical image. Tables 2, 3 and Figs. 10, 11 show that thresholding and binarization gives these metrics value near to 0. On the other hand, correlation coefficient value near to 1 denotes identical images. Table 4 also shows that image binarization and thresholding techniques gives better result. From the above results, we can conclude that image binarization and thresholding techniques is the best

Table 3 Relative area error

Image name	ACM	Thresholding	Watershed	Region grow
Image1	0.0700	0.0651	0.0584	0.0618
Image2	0.0387	0.0356	0.0444	0.0544
Image3	0.1851	0.1687	0.1806	0.1749
Image4	0.1246	0.1026	0.1244	0.1205
Image5	0.0873	0.0845	0.1007	0.1009
Image6	0.1603	0.1304	0.1314	0.1326
Image7	0.2290	0.2009	0.2406	0.2044
Image8	0.1733	0.1671	0.1955	0.1713

Fig. 11 Relative area error

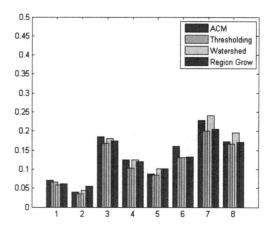

Table 4 Correlation coefficient

Image name	ACM	Thresholding	Watershed	Region grow
Image1	0.9992	0.9996	0.9994	0.9979
Image2	0.9993	0.9995	0.9990	0.9981
Image3	0.9994	0.9997	0.9989	0.9975
Image4	0.9990	0.9990	0.9985	0.9970
Image5	0.9992	0.9995	0.9975	0.9972
Image6	0.9993	0.9996	0.9980	0.9970
Image7	0.9995	0.9999	0.9987	0.9972
Image8	0.9994	0.9998	0.9986	0.9971

technique for automatic segmentation of lung nodule. We have used image binarization and thresholding techniques to extract all the required features of the lung nodules in Sect. 4.4 and we have predicted its risk using multiclass SVM.

4.4   Feature Extraction

The risk factor of a lung nodule depends on size, shape, growth rate, presence, and nature of calcification. In this study, we considered only size (radius) and shape of a lung nodule. Shape of a lung nodule depends on four parameters namely, size, aspect ratio, moments, and circularity [21]. We have selected a set of 15 candidate nodules from 40 detected solitary pulmonary nodule and tabulated their features in Table 5. In this study, we are not classified shape of lung nodule but we have considered all the values for classification procedure as shown in Table 6.

Table 5 Geometric features of nodules

Nodule	Maxl	Minl	Diameter	Eccentricity	Circularity	Aspect ratio
1	6.98	4.62	5.41	0.7496	0.9835	0.6618
2	13.33	8.19	10.02	0.7889	0.6948	0.6145
3	8.24	4.69	5.29	0.8221	0.6086	0.5693
4	7.25	4.71	5.41	0.7603	0.9021	0.6495
5	9.33	5.10	6.77	0.8372	0.9959	0.5469
6	6.49	6.05	6.07	0.3630	0.9689	0.9317
7	11.29	9.51	10.21	0.5380	0.9151	0.8429
8	4.69	3.27	3.74	0.6847	0.9597	0.7288
9	20.11	9.57	13.58	0.8793	0.7450	0.4761
10	27.30	16.62	18.40	0.7933	0.2364	0.6088

Table 6 Feature value of different shapes

shape	Size (mm)	Aspect ratio	Circularity	Second moments
Round	10.1	0.885	0.886	0.162
Lobulated	17.5	0.823	0.727	0.168
Polygonal	13.3	0.785	0.742	0.165
Tentacular	12.1	0.768	0.523	0.186
Spiculated	20.7	0.804	0.526	0.171
Ragged	21.1	0.860	0.514	0.170
Irregular	19.4	0.755	0.423	0.185

4.5 Result of Classification

According to LUNG-RADS [22], risk of a lung nodule can be categorized as benign, suspicious and malignant. Risk of a lung nodule depends on four features viz., size, shape, calcification, growth rate. In this study, we are considering only size and shape features of lung nodules. In [23], we find categorization SPN shape into seven groups viz., round, lobulated, tentacular, polygonal, speculated, ragged, and irregular; described in Table 6.

Nearly, 40 nodules are found from these images and are classified using multiclass SVM shown in Table 7.

Table 7 Categorization of malignancy

No. of nodules	Risk
4	Benign
6	Suspicious
30	Malignant
40	

4.6   Data Validation

In this section, we have calculated the specificity and sensitivity [1] of our algorithm to establish our algorithm.

1. Sensitivity

$$
\begin{aligned}
\text{Sensitivity} &= \frac{TP}{P} \\
&= \frac{TP}{TP + FN}
\end{aligned}
\tag{22}
$$

2. Specificity is a metric that excluded a condition correctly, i.e., proportion of healthy people not to have the disease, who will test negative for it. Mathematically, it can be defined as,

$$
\begin{aligned}
\text{Specificity} &= \frac{TN}{N} \\
&= \frac{TN}{TN + FP}
\end{aligned}
\tag{23}
$$

3. Accuracy

$$
\text{Accuracy} = \frac{TP + TN}{N + P}
\tag{24}
$$

where TP denotes true positive rates, means correctly identified, i.e., sick people correctly diagnosed as sick.

FP or false positives means correctly identified, i.e., healthy people incorrectly identified as sick.

TN or correctly rejected, i.e., healthy people correctly identified as healthy.

FN or false negative denotes incorrectly rejected, i.e., healthy people incorrectly identified as healthy.

P denotes the total number of population that have positive value and N denotes total number of population that have negative value.

Specificity and sensitivity of our algorithm are 75 and 72 % and FP are 11/scan.

5   Conclusion

Our experimental results illustrate an elaborate view of different image segmentation techniques that can automatically detect lung nodules from high-resolution CT images. We conclude that fast global minimization of active contour is the best image segmentation technique for lung nodule from thoracic CT images. Further, we have used the multiclass SVM for efficient risk prediction of lung nodule. In future, we will incorporate more segmentation techniques and also validate our test results with biopsy or cytology or bronchoscopy report of lung nodules.

Acknowledgments We are thankful to the Centre of Excellence in System Biology and Biomedical Engineering (TEQIP II), University of Calcutta for funding this project and Peerless Hospitex Hospital and Research Center Ltd. for providing their valuable lung cancer image database.

References

1. Sausa, J., Silva, A., Paiva, A., Nunes, R.: Methodology for automatic detection of lung nodules in computerized tomography images. Comput. Methods Progr. Biomed. **98**, 1–14 (2009). (Elsevier)
2. Gurcan, M.N., Berkman, S., Petrick, N., Chan, H.P., Kazerooni, E.A., Cascade, P.N., Hadjiiski, L.: Lung nodule detection on thoracic computed tomographyimages:preliminary evaluationofacomputer-aided diagnosis system. Med. Phys. **29**(11), 25522558 (2002)
3. Gao, T., Sun, X., Wang, Y., Nie, S.: A pulmonary nodules detection method using 3D template matching foundations of intelligent systems. Adv. Intell. Soft Comput. **122**, 625–633 (2012)
4. Ozekes, S., Camurcu, A.Y.: Automatic lung nodule detection using template matching advances in information systems. Lect. Notes Comput. Sci. **4243**, 247–253 (2006)
5. Netto, S., Silva, A., Nunes, R., Gattass, M.: Automatic segmentation of lung nodules with growing neural gas and support vector machine. Comput. Biol. Med. **42**, 11101121 (2012)
6. Suarez-Cuenca, J., Tahoces, P.G., Souto, M., Lado, M.: Application of the iris filter for automatic detection of pulmonary nodules on computed tomography images. Comput. Biol. Med. **39**, 921–933 (2009). (Elsevier)
7. Mukherjee, J., Chakrabarti, A., Shaikh, S.H., Kar, M.: Automatic detection and Classifications of Solitary Pulmonary Nodules from CT Images. IEEE Xplore (2014)
8. www.peerlesshospital.com
9. Otsu N.: A Threshold Selection Method from Gray-Level Histograms. IEEE Explore (1978)
10. Bression, X., Esedoglu, S., Vandergheynst, P., Thiran, J-P., Osher, S.: Fast global minimization of active contour/snake model. J. Math. Imaging Vis.
11. Rudin, L.I., Osher, S., Fatemi, E.: Nonlinear total variance based noise removal algorithms. Phys. D **10**(2), 259–268 (1992)
12. Mumford, J., Shah, J.: Optimal approximations of Piecewise smooth functions and associated variational problems. Commun. Pure Appl. Math. **42**, 577–685 (1989)
13. Caselles, V., Kimmel, R., Sapiro, G.: Geodesic active contours. Int. J. Comput. Vis. **22**(1), 61–79,1997
14. Kass, M., Witkin, A., Terzopoulos D., Snakes: Active contour models. Int. J. Comput. Vis. 321–331 (1987)
15. Chan, T., Vese, L.: Active contours without edges. IEEE Trans. Image Process. **10**(2), 266–277 (2001)
16. Gonzalez, R.C., Woods, R.E., Digital Image Processing, 3rd edn. Pearson Education (2009)
17. Chamasemani, F.F., Singh, Y.P.: Multi-class support vector machine (SVM) classifiers-an application in hypothyroid detection and classification. In Proceedings of Bio-Inspired Computing: Theories and Applications, pp. 351–356. IEEE (2011)
18. Dey, A., Shaikh, S.H., Saeed, K., Chaki, N.: Modified majority voting algorithm towards creating reference image for binarization. Adv. Comput. Netw. Inform. **1**, 221–227 (2014)
19. Roy S., Saha, S., Dey, A., Shaikh, S.H., Chaki N.z;; Performance evaluations of multiple image binarization algorithms using multiple metrices on standard image databases. Proc. Annu. Convention Comput. Soc. India **II**, 349–360 (2014)
20. Mukherjee, J., Kundu, R., Chakrabarti, A.: Variability of cobb angle measurement from digital ray image based on different denoising techniques. Int. J. Biomed. Eng. Technol. **16**(2), 113–134

21. Mughal, M.N., Karim, W.: Early Lung Cancer Detection by Classification by Classifying Chest CT Images: A Survey. IEEE (2004)
22. Manos, D., Sely, J.M., Taylor, J.T., Borgaonkar, J., Roberts, H.C., Mayo, J.R.: The lung reporting and data system (LU-RADS): a proposal for computed tomography screeing. Can Assoc. Radiol. J. **65**, 121–134 (2014). (Elsevier)
23. Iwano, S., Nakamura, T., Kamioka, Y., Ishigaki, T.: Computer-aided diagnosis: a shape classification of pulmonary nodules imaged by high resolution CT Elsevier. Comput. Med. Imaging Graph. **29**(2005), 565–570 (2005)

Part III
Networking and Cloud Computing

Safe Cloud: Secure and Usable Authentication Framework for Cloud Environment

Binu Sumitra, Pethuru Raj and M. Misbahuddin

Abstract Cloud computing an emerging computing model having its roots in grid and utility computing is gaining increasing attention of both the industry and laymen. The ready availability of storage, compute, and infrastructure services provides a potentially attractive option for business enterprises to process and store data without investing on computing infrastructure. The attractions of Cloud are accompanied by many concerns among which Data Security is the one that requires immediate attention. Strong user authentication mechanisms which prevent illegal access to Cloud services and resources are one of the core requirements to ensure secure access. This paper proposes a user authentication framework for Cloud which facilitates authentication by individual service providers as well as by a third party identity provider. The proposed two-factor authentication protocols uses password as the first factor and a Smart card or Mobile Phone as the second factor. The protocols are resistant to various known security attacks.

Keywords Cloud computing · Authentication issues · Single sign-on · SAML · Mobile authentication

B. Sumitra (✉)
Christ University, Bangalore, India
e-mail: sumitrabinu@gmail.com

P. Raj
IBM Cloud Global Center of Excellence, IBM India Pvt. Ltd., Bangalore, India
e-mail: peterindia@gmail.com

M. Misbahuddin
C-DAC, Electronic-City, Bangalore, India
e-mail: mdmisbahuddin@gmail.com

© Springer India 2016   183
R. Chaki et al. (eds.), *Advanced Computing and Systems for Security*,
Advances in Intelligent Systems and Computing 395,
DOI 10.1007/978-81-322-2650-5_12

1 Introduction

The advent of Web 2.0 has contributed to an exponential growth in the users of Internet and related technologies. Cloud computing, an Internet-based distributed computing model offering computing resources as a service, is a fast growing technology slowly being embraced both by corporate sector as well as by laymen. With the advancements in IT, this technology has evolved through a number of different services such as software applications (SaaS), computing platform, (PaaS), and infrastructure (IaaS). Thus Cloud computing refers to both the applications delivered as services over the Internet and the hardware and system software in the data centers that provide those services [1]. The primary objective of Cloud computing is to provide great flexibility to users, by allowing the users to process, store and access their data using Cloud services, anytime, anywhere using the Internet without investing on Computing Infrastructure.

The fast growing utility-based Cloud computing technology offers many advantages such as resource sharing dynamic scalability, rapid elasticity, efficient software/platform/infrastructure utilization, and many more [2]. However, this technology has a lot of concerns including performance, resiliency, interoperability, data migration and transition from legacy systems, preventing the whole hearted adoption of Cloud services. Among the many issues, one of the most relevant is security, which involves virtualization security, distributed computing, application security, identity management, access control and authentication [3–5]. Furthermore in [6, 7] authors have pointed out that the identity and access control management is a core requirement for Cloud computing. Hence, strong user authentication which thwarts illegal access of Cloud servers becomes the fundamental requirement in the Cloud computing environment.

Over the last few years, significant research has happened in the security-related areas of Cloud computing with the objective of arriving at mechanisms that provide adequate security to Cloud environment and its users [2, 8–10]. However, these existing security mechanisms are susceptible to certain security attacks, when examined from the perspective of practical implementation. Many existing public Clouds such as Amazon Web Services, Dropbox, Salesforce.com, and Google App Engine, etc., have been victimized to various security attacks [11–14]. Hence it is possible for illegal users to exploit these security flaws and either steal secret information or disturb the normal operation of Cloud service providers by launching various attacks. This points out to the requirement for securing Cloud with a strong user authentication mechanism that prevents illegal users from performing various nefarious activities in the Cloud.

Authenticating the identity of remote users is a preliminary requirement in a public Cloud environment before they can access a secure resource. Service providers (SP) should ensure that only authorized users are accessing services provided by the application system and password authentication is one of the simplest and most convenient user authentication mechanism. Unfortunately, users tend to choose low entropy passwords which are easy to remember rendering the authentication system susceptible to various attacks. Strong authentication mechanisms address this issue by authenticating users based on a combination of two or more factors, viz., what you know, what you have, and what you are.

Taking into consideration, the security issues faced by Cloud, the discussed work proposes a strong and reliable two-factor user authentication framework for Cloud environment. The rest of the paper is organized as follows. Section 2 discusses the related work. Section 3 discusses novelty of our contribution, Sect. 4 explains the authentication framework, Sects. 5 and 6 elaborate the authentication architectures and protocols respectively and Sect. 7 concludes the work done.

2 Related Work

Choudhary et al. proposed a user authentication framework for Cloud environment [2] using password, Smart Card, and out-of-band (OOB) authentication token. Cloud environment comprises of multiple service providing servers and users may access services from servers belonging to different domains. The single sign-on (SSO) functionality which provides convenient user authentication in such a scenario was not considered by Choudhary et al. In 2013 Jiang [10] proved that the scheme [2] was prone to masquerade user attack, the OOB attack, and the password change flaw and proposed a modified scheme. The scheme addresses the security issues of [2], but uses time stamps which can lead to time synchronization problems. The protocol [15] stores a variant of the user password in the server which makes it susceptible to stolen verifier attack.

Sanjeet et al. [16] proposed a user authentication scheme using symmetric keys for Cloud services. The protocol uses a one-time token which is sent to the registered users e-mail ID. This scheme requires the user to login to two accounts during the authentication process which may cause user inconvenience. As in the case of [2], the authentication schemes proposed by Rui Jiang and Sanjeet et al. does not provide the SSO functionality which enhances user convenience in a multi-server Cloud environment.

3 Novelty of Our Contribution

Primary objective of the work is to design an authentication framework for cloud services encompassing authentication models and authentication protocols, which can be used by two categories of organizations, namely collaborative organizations and financial institutions. The two-factor authentication protocols are designed to overcome the limitations of currently prevalent mechanisms and be capable of operating in a traditional environment as well as in a Smart environment. The authentication model addresses the issues related to storing passwords at the cloud service provider's end.

The proposed lightweight authentication protocols are designed to have minimum processing overhead and to be resistant to the common attacks on authentication.

4 User Authentication Framework for Cloud

This section provides an overview of the proposed authentication framework for Cloud and discusses some approaches to address its components. The overall authentication framework and the key components required to integrate and provide services in a secure manner are depicted in Fig. 1. The proposed framework comprises of three major entities, viz., the users, the cloud brokers (CBs), and the cloud service providers (CSPs). The CBs act as an identity provider (IdP) and as an intermediary between the users and CSPs. They work with the users to understand the work processes, provisioning needs, budgeting, data management requirements, etc. The CBs then discover services from different CSPs or other brokers, carry out negotiations, integrate the services to form a group of collaborating services and recommend the concerned CSPs to the user. Each CB has components that are

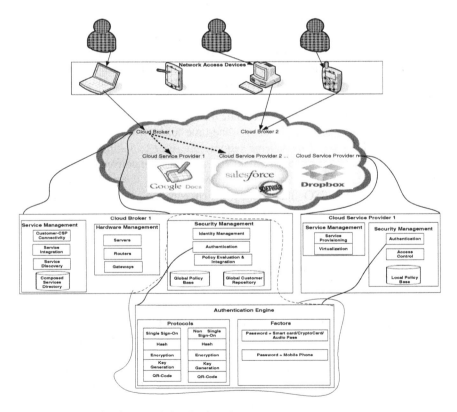

Fig. 1 Authentication framework for cloud environment

responsible for ensuring security and establishing trust between local provider domains and between the providers and users as well as generating global policies. In a Cloud environment, users should have some level of trust on each other.

4.1 Components of the Framework

This section details the functionality of the various components of Cloud Brokers and the CSPs of the framework.

Service Management This component of CB is responsible for secure service discovery by interacting with the CSPs, integrating the services, composing new desirable services, and establishing the link between the user and the CSPs [17]. Discovery of services, integrating the services, and connecting the users and CSPs are done by the service discovery module, service composition module and the customer–CSP connectivity module respectively. The Service Management component of the CSP is responsible for provisioning the services to the user. To support multi-tenancy and resource sharing, the CSP uses Virtualization technology.

Security Management The Security Management component comprises of the modules whose functionality aids in enforcing security and trust. The identity management module of CB is responsible for issuing and managing the identification credentials of the users registered with the CB. The authentication module is responsible for authenticating users and CSPs based on the credentials. The authentication module of CB executes a two-factor authentication protocol and uses SAML to provide user convenience through Single Sign-On functionality. The CSPs may have conflicting interests regarding the policies they adopt to provide services to their users and this may be a matter of concern when multiple CSPs collaborate to provide a customized service. Specification frameworks are needed to ensure that the cross domain accesses are properly specified and enforced [17]. SAML, XACML, and WS standards are viable solutions toward these needs and the proposed framework uses to address this requirement. The policy evaluation and integration module examines the policies of various CSPs whose services need to be integrated. The module then addresses security challenges such as semantic heterogeneity, secure interoperability, and integrate access policies of different CSPs and define global policies to accommodate the requirements of all the collaborating CSPs. These global policies are available in the global policy repository. The security management component also includes a global customer repository that stores the details of the registered CSPs and users.

Authentication Engine The functionality of the authentication engine is accessed by the authentication module of both the CBs and the CSPs to authenticate the users prior to providing access to the services provided by the CSPs. The proposed authentication protocols verify user authenticity by a two-factor authentication mechanism, and do not require the authentication server to maintain a verification table. The scheme uses password as the first factor and a Smart Card/Crypto Card/Audio Pass/Mobile Phone as the second factor. The advantage of using an Audio Pass is that it can be used with smart phones and hence do not require an additional device (Card reader) to read the stored data as in the case of Smart Cards/Crypto Cards. The paper proposes two different authentication architectures, one in which the IdP/CB authenticates the users prior to accessing the services of CSP and thus provides a SSO facility. In the second authentication the users are authenticated by the CSPs whose service they wish to access. In both the architectures both the users and the CSPs need to first register with the IdP/CB. The authentication protocols supported by both the architectures provide two-factor authentication by using password as the first factor and Smart Card/Crypto Card/Audio Pass/Mobile Phone as the second authentication factor. Smart Card/Crypto Card/Audio Pass/USB Token is primarily meant to be used in the case of those departments/ organizations/users for which security is the top priority and hence the additional hardware cost and the extra inconvenience is acceptable. Mobile phone as a second authentication factor is targeting on those departments/organizations who would like to ensure security by making use of a commodity that the user already owns rather than burdening the user with an additional hardware [18].

5 Authentication Architectures

The following sections give a brief overview of the authentication architectures proposed in this work.

5.1 Broker-Based Authentication Architecture

The Coca-Cola company is collaborating with Heinz to use their bottling factory to make PET bottles using 100 % plant-derived materials and plant residues. To effectively reuse the used bottles, the Coca-Cola Company and furniture company Emeco have entered into a collaboration to make Emeco 111 navy chair, a chair made of 111 recycled bottles. Similarly, Biotherm, a skin care company and the automobile manufacture Renault has collaborated to invent the new concept in cars, 'The Spa Car.' These collaborative organizations can offer their services via the public cloud which offers the advantages of resource sharing, standardization of operations, increased reusability, reduced capital expenditure, etc. These collaborative organizations will be having a common customer base who would like to access the services and information from all these organizations. When these individual organizations offer their services from a cloud environment, each of them will have their own applications server to provide the services and the information. In a conventional environment, a customer who wants to access services of all these collaborating organizations will be required to create individual accounts with each

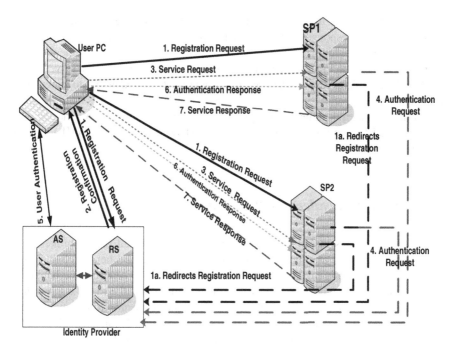

Fig. 2 Registration and authentication process flow in broker-based authentication architecture

service provider. This contributes to multiple accounts, multiple passwords and multiple authentication to access multiple services. Nevertheless these organizations would prefer to provide their customers with a user-friendly procedure that enables them to access information from all the collaborating partners with ease. The work proposes the use of a single account to access the services of multiple service providers and a user-friendly login process that allows the user to authenticate once and access multiple services during a single login session, termed as single sign-on. To support single account and single sign-on, we propose a broker-based architecture comprising of a centralized registration authority alias identity provider (IdP) and the multiple service providers. All the service providers and the users accessing the services of these service providers should be registered with the registration server of the centralized registration authority alias IdP. The users can register once and create an account at the IdP to get the services of all the registered collaborating service providers. Once registration is done user is issued an authentication token such as a Smart Card/Crypto Card/Audio Pass or he is given an option to download a secret file into his Mobile Phone if he is using Mobile Phone as the second authentication factor. To facilitate single sign-on, the authentication of users is also done by the authentication module of IdP who is trusted by the collaborating service providers. During login process, the service providers will redirect the users to the IdP who will authenticate the users using the proposed two-factor authentication protocol and send the response (token/assertion) to the service provider.

Thus user can access the services seamlessly and the services providers having handed over the responsibility of authenticating their customers to the IdP can concentrate on providing their core services. The registration and authentication process is depicted in Fig. 2. The registration phase includes registration server (RS), CSP, and users'. The RS and AS are in the same trusted domain and together they provide the functionality of the identity provider (IdP/CB). The CSPs and IdP work in a trust-based environment. The proposed architecture inherits all the desired features of a MSE and uses Security Assertion Markup Language (SAML) to provide user convenience through SSO [19].

5.2 Direct Authentication-Based Architecture

Financial institutions including commercial banks, investment banks, brokerages, insurance companies, etc., deal with highly sensitive data and hence reliable customer authentication is imperative for engaging in any form of electronic banking. These organizations when they move into the cloud would like to be assured that their data and information stored in the cloud is completely secure from unauthorized access. A strong and effective authentication system can help financial institutions to reduce fraud, to enforce anti-money laundering practices, and detect and reduce identity theft. The risk of engaging in business with unauthorized individual in a money-issuing environment could result in financial loss and reputation damage through fraud, disclosure of confidential information, corruption of data, etc. Considering these risks, these financial institutions when they operate from the cloud will not be willing to trust anybody regarding the authenticity of the users attempting to access their data and services and would prefer to have an internal mechanism for authentication. To address such a scenario, the work proposes a direct authentication

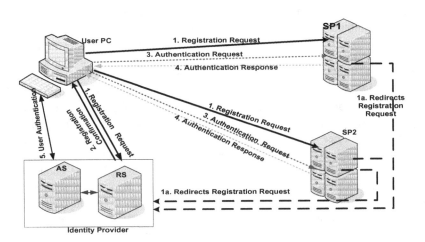

Fig. 3 Registration and authentication process flow in direct authentication-based architecture

architecture as shown in Fig. 3, where we place an authentication server in front of each service providing server belonging to the category of financial institutions. In this architecture, the users register themselves at the Identity Provider who issues them with an authentication token as in the case of broker-based architecture. After registering, a user who wants to access the services would attempt to login to the server of the financial institution and these servers will be independently authenticating their users using the proposed two-factor authentication protocol.

6 Authentication Protocols

The authentication protocols of the proposed work are categorized into two broad categories, viz., protocol for broker-based authentication architecture and protocol for Direct Authentication-based architecture. Again the two broad categories are subdivided into the sub categories (i) Protocol for broker-based authentication using password as the first factor and Smart Card/Crypto Card/Audio Pass as the second factor (ii) Protocol for broker-based authentication using mobile phone as the second factor (iii) Protocol for direct authentication using password as the first factor and Smart Card/Crypto Card/Audio Pass as the second factor (iv) Protocol for direct authentication using mobile phone as the second factor. This section discusses the various phases of the first three categories.

6.1 Protocol for Broker-Based Authentication Using Password as the First Factor and Smart Card as the Second Factor

Phases of the Protocol The proposed scheme consists of four phases, viz., Initialization phase, User registration phase, Login and Authentication phase, and Password change phase. The notations used in the protocol are listed in Table 1.

Table 1 Notations used in broker-based authentication using smart cards

AP, SC	Audio pass, smart card
U_a, IdP, SP	User 'a', identity provider, service provider
ID_a, P_a	Identity, password of user U_a
SID, y	Server ID of IdP, secret key of IdP
G	Additive cyclic group of prime order
g_0	Generator of additive cyclic group
r	Random number generated by audi pass unique to each session
$h(.)$, $\oplus$, $\parallel$	Hash function, XOR operation, concatenation operation
$\Rightarrow$	Secure communication channel

Initialization Phase During this phase, User U_a generates a finite additive cyclic group 'G' of prime order 'n' and selects an element 'g_0' from the group. 'g_0' is one of the generators of 'G' and is used by U_a to modify the password to be used for secure user registration and authentication.

User Registration Phase If user wants to register for the services of a Service Provider SP, the user U_a clicks the 'Create Account' link at SPs web site. SP redirects U_a to the registration page of the IdP. IdP prompts U_a to submit the Identity and Password of the user. U_a chooses her identity ID_a and Password P_a and the phase proceeds as illustrated in Fig. 4, which can be explained as follows.

UR1: U_a Computes $b = h(P_a)$, $k = g_0^b$ and submits $h(ID_a)$, k to IdP through a secure channel. IdP checks whether the submitted $h(ID_a)$ already exists in its user table and if so prompts U_a to submit a new ID, otherwise IdP proceeds as follows:

IdP computes $E_i = B_i \oplus h$ (SID$\|h(y)$) where 'y' is a secret key of IdP and $h(.)$ is a one way hash function. $B_i = h(h(ID_a)\|h(SID))$; $J_i = h(SID\|h(y)) \oplus k$; $C_i = h(h(ID_a) \| h(SID\|h(y))\| k)$;

UR2: IdP personalizes the smart card (SC) with the parameters C_i, E_i, J_i, $h(.)$. IdP sends the SC to U_a via a secure channel.

UR3: On receiving the device, U_a stores g_0 into the Audio Pass/Smart Card/USB token which now contains $\{C_i, E_i, J_i, h(.), g_0\}$.

Login and Authentication Phase This section discusses the Cloud Service Provider initiated authentication and Fig. 5 gives a pictorial representation of the same.

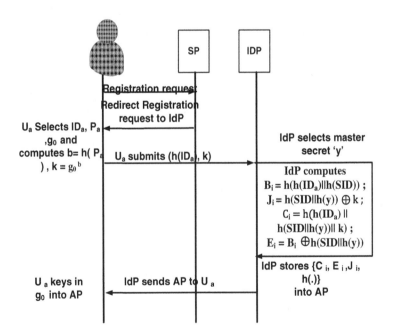

Fig. 4 Registration phase of broker-based authentication protocol using smart card/audio pass

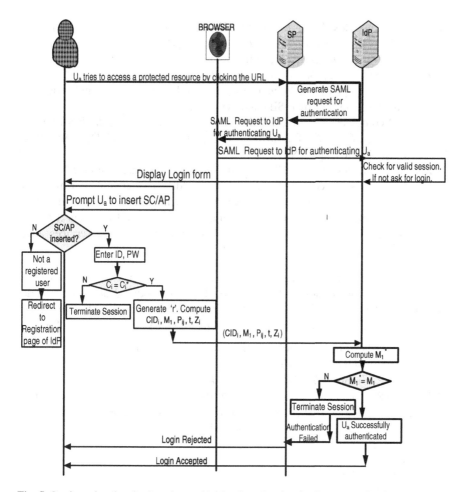

Fig. 5 Login and authentication phase of broker-based authentication protocol using smart card

Whenever a registered user wants to login to access the services of the Service Provider 'SP', she attaches the Smart Card or Audio Pass token to the system and proceeds as follows:

UL1: U_a requests for login to the SP

UL2: SP checks for an existing session with U_a and if there is no valid session, SP redirects U_a to IdP with a SAML authentication request

UL3: U_a keys in her ID_a and P_a

UL4: SC/AP computes, $b = h(P_a)$, $k^* = g_0^b$

UL5: SC/AP computes $h(SID|h(y))^* = J_i \oplus k^*$

UL6: SC/AP computes $C_i^* = h(h(ID_a) \| h(SID\|h(y))^* \| k^*)$ and compares with C_i stored in the AP. If invalid, AP terminates the session. Otherwise generates the login message as follows:

UL7: AP generates a random number 'r' and computes nonce $n_1 = g_0^r$

UL8:   AP computes $P_{ij} = E_i \oplus h(h(SID\|h(y))\|n_1)$; $B_i = E_i \oplus h(SID\|h(y))$; $CID_i = C_i \oplus h(B_i\|n_1\|SID)$; $M_1 = h(P_{ij}\,|C_i\,\|B_i\|n_1)$; $t = g_0 \oplus h(SID\|h(y))$; $Z_i = (r - CID_i) \oplus h(SID\|h(y))$

UL9:   AP sends $(CID_i, M_1, P_{ij}, t, Z_i)$ to IdP

UL10:   Upon receipt of the login message the IdP performs the authentication process using her own SID and $h(y)$ values

UL11:   IdP computes, $r = (Z_i + CID_i) \oplus h(SID\|h(y))$; $g_0 = t \oplus h(SID\|h(y))$; $n_1^* = g_0^r$, $E_i = P_{ij} \oplus h(h(SID\|h(y))\|n_1)$; $B_i^* = E_i \oplus h(SID\|h(y))$; $C_i^* = CID_i \oplus h(B_i^*\|n_1^*\|SID)$;

UL12:   IdP computes $M_1^* = h(P_{ij}\,\|C_i^*\,\|B_i^*\|n_1^*)$ and compares with the M_1 in the login message received from U_a. If valid, IdP considers the authentication as successful and creates a SAML authentication response message containing the result of the authentication process and redirects it to the SP. The Service Provider permits or denies access to the services after verifying the response from the IdP.

Password Change Phase A registered user can change her password by selecting the change password option and the password can be modified at the client side without the intervention of IdP and SP. The change password request is processed only if the keyed in ID and password are valid. This phase proceeds as follows:

UP1: U_a attaches his SC or AP into the system and keys in his ID_a and P_a

UP2:   SC/AP computes, $b = h\,(P_a)$, $k = g_0^b$

UP3:   SC/AP computes $h(SID\|h(y)) = J_i \oplus k$

UP4:   SC/AP computes $C_i^* = h(h\,(ID_a)\,\|\,h(SID\|h(y))^*\|\,k^*)$ and compares with C_i stored in the SC/AP. If invalid, SC/AP terminates the session. Otherwise prompts U_a to enter the new password

UP5: U_a enters P_{anew}

UP6:   SC/AP computes $b_{new} = h(P_{anew})$; $k_{new} = g_0^{bnew}$; $J_{inew} = J_i \oplus k \oplus k_{new}$; $C_{inew} = h(h(ID_a)\,\|\,(J_i \oplus k)\|\,k_{new})$;

UP7:   SC/AP replaces C_i and J_i in the AP/SC with C_{inew} and J_{inew} respectively.

Security Analysis of the Protocol This section analyzes the security of the proposed protocol against various attacks.

Security Against Guessing Attack The proposed protocol is secure against guessing attack as it is impossible within polynomial time, for an adversary to retrieve user's password P_a or IdP's secret key 'from the intercepted parameters $(CID_i, M_1, P_{ij}, t, Z_i)$.'

Security Against Malicious Insider Attack In the proposed scheme, user submits $k = g_0^{h(Pa)}$ to IdP rather than the plain text form of the password. This guards the password from being revealed to IdP and hence even if the user uses the same password to login to other servers, her credentials will not be susceptible to insider attack.

Security Against Replay Attack A replay attack is launched by the adversary by capturing a message exchanged between the Client and Server and replaying at a later point in time. The scheme is resistant to replay attack since nonce values used to in each authentication message is unique and varies for each session. Hence the IdP will be able to identify a replayed login message (CID_i, M_1, P_{ij}, t, Z_i) by checking the freshness of nonce, $n_1 = g_0^r$ where 'r' is a random number generated by user and is unique to a session.

Security Against Stolen Verifier Attack The fact that the proposed scheme does not require the Server to maintain a verifier/password table makes it resistant to Stolen Verifier attack.

Security Against User Impersonation Attack If an adversary attempts to impersonate a valid user, he should be able to forge a valid login request on behalf of the user. In the proposed scheme if an adversary intercepts the login message (CID_i, M_1, P_{ij}, t, Z_i) and attempts to generate a similar message, he will fail since the value of nonce 'n_1' as well as the server's secret key 'y' is unknown to him.

Security Against Denial-of-Service (DoS) Attack A DoS attack can be launched by an adversary by creating invalid login request messages and bombarding the server with the same. This attack can also be launched by an adversary who has got control over the server and is able to modify the user information stored in the server's database which in turn prevents the valid user from accessing the resources.

The first scenario will not work in the case of the proposed scheme, since it is impossible for the adversary to create valid login request messages without knowing the password. The validity of the password is checked at the client side before creating a login request. The second scenario is also not applicable in the proposed scheme since the server does not maintain a verifier/password table.

Security Against Smart Card Lost Attack If the adversary steals the Smart Card containing the parameters (C_i, E_i, J_i, $h(.)$, g_0), he can neither retrieve the user's password nor the IdP's master secret 'y' from the stored value. To extract the password from $k = g_0^{h(Pa)}$, the adversary needs to solve the discrete logarithm problem. Again the password is used in the hashed form which is irreversible. Also, retrieving the IdP's master secret, 'y' is not possible without knowing the password of the user which is unknown to the adversary.

6.2 Protocol for Broker-Based Authentication Using Password as the First Factor and Mobile Phone as the Second Factor

This protocol consists of a registration phase, login and authentication phase, and a password change phase as elaborated in the following subsections. Here, it is assumed that the CSP who wants to be a part of the framework is registered with the IdP.

Phases of the Proposed Protocol The proposed scheme consists of three phases, viz., User registration phase, Login and Authentication phase, and Password change phase. The notations used are listed in Table 2.

Table 2 Notations used in broker-based authentication using mobile phone

ID_a, P_a	Identity, password of user U_a
s, rand	Secret key, random number of RS
r, Key_a, V_a, m	Random number generated by AS unique to each session, master secret key of user, parameter used for verifying the password, parameter used for changing the password
$h(.)$, $\oplus$, $\parallel$	Hash function, XOR operation, concatenation operation
$\Rightarrow$	Secure communication channel

Registration Phase The registration process illustrated in Fig. 6 can be explained as follows:

UR1. The user U_a clicks the 'Create Account' link at SPs web site. SP redirects U_a to the registration page of the IdP. RS prompts U_a to submit her identity ID_a and PW_a.

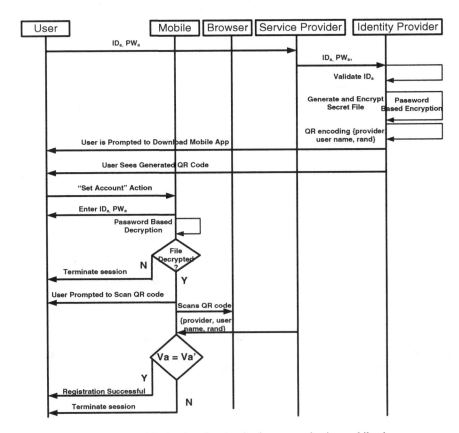

Fig. 6 Registration phase of broker-based authentication protocol using mobile phones

RS checks whether ID_a already exists in its user table. If so U_a is prompted to select a new ID_a.

UR2. RS generates a secret key 'S' and selects a random number 'rand'. RS creates a file containing the authentication parameters V_a, Key_a, m and the file is encrypted using password of U_a and a salt value. 'V_a' is used only during the registration and installation of mobile application. 'Key_a' is used during the authentication phase and 'm' is used during the password change phase. The values of V_a, Key_a, m are generated by performing hash and XOR operations on ID_a, PW_a, S, and rand.

UR3. The RS generates a QR code embedding 'rand', Service Provider Name and User Name and the QR code will be displayed on the web application screen and the user will be prompted to download the mobile application. Along with the app, the secret file will also be imported and stored in a safe location within the user's phone in the form of a mobile token. When the user touches the register button in the mobile app, the user will be prompted to enter his ID_a, PW_a. The app attempts to decrypt the file using password given as input by the user and the salt value attached to the file.

UR4. If the decryption is successful, the app invokes the scanning application, and the user can scan the code. The mobile app retrieves 'rand' from QR code, calculates V_a' using password and 'rand'. V_a' is compared with V_a stored in the mobile token and if equal, the registration process is considered successful and the account is created. RS stores the user identity in its user table.

Login and Authentication Phase As shown in Fig. 7, the user via his browser attempts to access a protected resource of a Service Provider (SP). It is assumed that, the browser at this point does not have an established session with the SP. On receiving the request from the user, SP redirects the user to the login page of IdP and requests IdP to authenticate the user. The authentication request contains the URL of the SP who initiated the request. Also the request should contain the URL to which the response should be sent. IdP checks for a valid session with the browser. If there is no existing session between the browser and the IdP, then IdP generates a login session and authenticates the user by executing the authentication phase, as illustrated in Fig. 7. The procedure can be explained as follows:

UA1. Authentication server (AS) displays the login page and prompts the user to enter user's identity (ID_a) and Password (PW_a). AS calculates Key_a and a challenge 'B' using server's secret key 'S', random number 'r' and the ID and PW of U_a. The random number r, B is send to the user via a secure communication channel. The mobile app computes B' using Key_a and the received random number 'r', where Key_a is the master secret key stored in the mobile token within the phone.

UA2. Mobile app checks whether B' = Challenge B, received from AS. If so, mobile app considers the message as being received from an authenticated source. Mobile app sends $K = HMAC(Key_a, B')$ to IdP. IdP on receiving the message K, computes $K' = HMAC(Key_a, B)$ and checks whether it is equal to the received K. If equal IdP considers the user as authenticated and that the integrity of message is maintained.

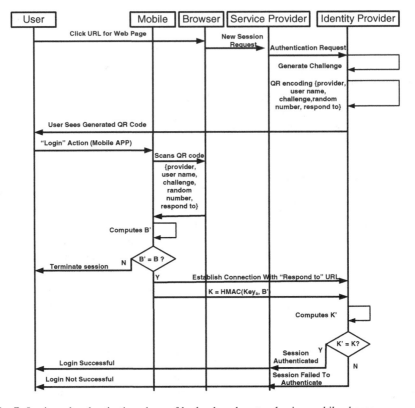

Fig. 7 Login and authentication phase of broker-based protocol using mobile phones

UA3. IdP creates an assertion token containing the result of the authentication process. IdP sends the token to the SP. The SP verifies the token issued by the IdP. It is assumed that the IdP and SP works in a trust-based environment. If the authentication is successful and the token is valid then SP notifies the user's browser of a successful login. Otherwise the login request is rejected.

Password Change Phase The password change phase is invoked when the user wishes to change his password without the intervention of the IdP or the SP is carried out as follows:

UP1. User enters his identity (ID_a) and Password (PW_a) and executes the 'Password Change' request. The mobile app computes $m' = Key_a \oplus h(ID_a \parallel PW_a)$ and checks whether it is equal to stored 'm'. If equal, the mobile app prompts the user to enter the new password '$PW_{a\ new}$'. Otherwise the 'password change' request is rejected.

UP2. The app calculates $h(ID_a \parallel h(s)) = Key_a \oplus h(PW_a)$. Then the app computes $Key_{a\ new} = h(PW_{a\ new}) \oplus Key_a \oplus h(PW_a)$ and $m_{new} = Key_{a\ new} \oplus h(ID_a \parallel PW_{a\ new})$ and replaces the existing values in the file with the new values.

6.3 Protocol for Direct Authentication Using Smart Card/Crypto Card/Audio Pass as the Second Factor

Phases of the Protocol The proposed scheme consists of four phases, viz., Initialization, Registration, Login and Mutual Authentication, and the Password change phase as explained in the following paragraphs. The notations used are listed in Table 3.

Initialization Phase RA selects two prime numbers p and q. RA computes $n = p *q$ where n is public and p and q are secrets. RA chooses a master secret key 'x' and a secret number 'd'. The keys of RA should be generated and managed using standard hardware security management (HSM) module.

Registration Phase In this scheme every server and user has to first register with RA. This phase is divided into two subphases: (1) Service provider integration with RA and (2) User registration phase.

Service Provider Integration with RA S_j selects its identity SID_j and submits the registration request to RA. RA after verifying S_j, computes SS_j and sends $\{SS_j, h(d), n\}$ to S_j via a secure channel. It is assumed that RA and S_j communicates using signed messages and thus works in a trusted environment.

User Registration Phase The registration process as shown in Fig. 8 can be explained as follows. U_i chooses his identity ID_i, a PIN number 'S' and submits $\{h(ID_i), S\}$ to RA. RA generates a random number 'r' and computes the Initial secret of U_i as I_i. RA selects a random password PW_i for U_i and computes $V_i, B_i, C_i, D_i, E_i, Y_i$. RA stores $\{C_i, D_i, E_i, Y_i, n, h(.)\}$ into the smart card which is issued to U_i via a secure channel. RA sends PW_i to U_i via an SMS and updates the user table.

Login and Authentication Phase Login and authentication phase runs independently on each SP and is executed once for each session. When U_i wants to access the services of S_j, he carries out the login process, which can be explained as follows.

U_i requests for a protected resource of S_j by typing the URL or by clicking the link for logging in. The login page is displayed. User is prompted to swipe the smart card and click 'proceed' button. The rest of this phase will be executed on the basis

Table 3 Notations used in direct authentication using smart cards

RA, SP, SC	Registration authority, service provider, smart card
U_i, S_j, SID_j	ith user, jth service provider, ID of jth service provider
ID_i, PW_i, S	Unique identification of U_i, password of user U_i, pin number chosen by U_i
p, q, n, x, d	Prime numbers, public parameter, master secret, secret number of RA
I_i	Initial secret unique to U_i generated by RA
r, k, N_1, N_2	Random number generated by RA, SC respectively, nonce values
$h(.), \oplus, \|, \Rightarrow$	Hash function, XOR, concatenation, secure communication channel

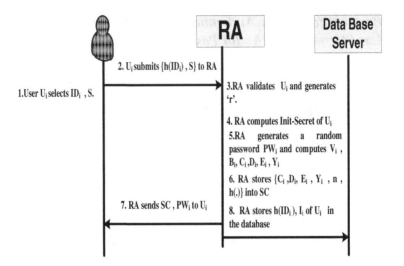

Fig. 8 Registration phase of direct authentication-based protocol using smart card

of the following three possible scenarios. **(i) User is not a registered user**: U_i has not registered with RA and hence does not have a smart card to swipe. In this case the user is directed to the registration page of RA and the registration phase is carried out **(ii) User has entered the smart card but is an invalid user**: U_i inserts his smart card into the reader. U_i is prompted to enter his ID_i and PW_i. SC computes V_i', I_i', D_i' and compares D_i' with D_i stored in the smart card. They are not equal and SC terminates the session. **(iii) User has entered the smart card and is a valid user:** U_i inserts his smart card into the reader. U_i is prompted to enter his ID_i and PW_i. SC computes V_i', I_i', D_i' and compares D_i' with D_i stored in the SC. If they are equal, the system checks whether U_i is accessing any SP for the first time. If so the system prompts U_i to change his password and the values in the smart card are modified accordingly. In this manner, the true password of the user is neither stored at the RA or nor does it travel across the network. Then SC proceeds to generate the login request message.

SC generates a random number 'k' and computes the parameters N_1, B_i, K_i, Q_j, P_{ij}, CID_i, Z_i. SC sends $\{h(ID_i), P_{ij}, CID_i, M_1, Z_i\}$. On receiving the request from U_i, server S_j checks whether there is an entry corresponding to $h(ID_i)$ in its database. If it exists then the corresponding init-secret, I_i is retrieved from the local database. Otherwise a request for init-secret, is sent to RA along with the $h(ID_i)$. RA retrieves the same from its database and sends init-secret I_i to S_j. S_j updates its user table with username and Init-Secret. Server S_j computes k and N_1. S_j checks the freshness of the nonce N_1 and computes Q_j', Y_i, E_i, B_i', D_i', P_{ij}, CID_i, M_1'. S_j compares M_1' with the M_1 in the login request. If the equality does not hold S_j rejects the request. Otherwise S_j generates a nonce N_2 and computes M_2, E_i', M_3. S_j sends $\{M_2, M_3\}$ to U_i. On receiving $\{M_2, M_3\}$, U_i computes N_2 and M_3^*. U_i compares M_3^* with M_3 in the response message from S_j. If equal U_i authenticates S_j successfully and generates a mutual authentication message M_4 and sends $\{h(N_2 + 1), M_4\}$ to S_j. If M_3^* is not equal to M_3, U_i terminates the session. On receiving $\{h(N_2 + 1), M_4\}$, S_j computes $h(N_2 + 1)$ and checks the freshness of the nonce. Then S_j computes M_4' and compares with the received M_4. If equal then mutual authentication holds. Otherwise S_j terminates the session. After mutual authentication, both U_i and S_j compute the session key $SKey_{ij}$.

7 Conclusion and Future Work

This work proposes a user authentication framework for Cloud environment, which attempts to address the authentication issues in Cloud. The authentication architecture and protocols of the proposed scheme addresses the issue of maintaining multiple authentication credentials in a multi-server environment by adopting SAML Single Sign-on functionality. The framework also caters to the requirements of those departments/organizations that prefer to independently authenticate their customers and this is made feasible by providing the option of executing the authentication protocol by the CSPs. The authentication protocols use two-factor authentication where password is the first factor and Smart Card/Crypto Card/Audio Pass/USB Token/Mobile Phone is the second factor. Security analysis of the protocols has been done and the protocols are resistant to user impersonation attack, server impersonation attack, replay attack, insider attack, parallel session attack, smart card lost attack, and the like.

References

1. Armbrust, M., Fox, A., Griffith, R., Joseph, A.D., Katz, R., Konwinski, A., Lee, G., Patterson, D., Rabkin, A., Stoica, I., Zaharia, M.: Above the clouds: a berkely view of cloud computing. Technical report no. UCB/EECS-2009-28. http://www.eecs.berkeley.edu/Pubs/TechRpts/2009/EECS-2009-28.pdf
2. Amlan, J.C., Pradeep, K., Mangal, S., Hyota, E.L., Hoon-Jue-Lee: A strong user authentication framework for cloud computing. IEEE Asia-Pacific Services Computing Conference (2011)
3. Chakraborty, R., Ramireddy, S., Raghu, T.S., Rao, H.R.: The information assurance practices of cloud computing vendors. IT Prof. 12, 29–37 (2010)
4. Miller, H.G., Veiga, J.: Cloud computing: will commodity services benefit users long term? IT Prof. 11, 29–39 (2010)
5. Blumenthal, M.S.: Hide and seek in the cloud. IEEE Secur. Priv. 8, 29–37 (2010)
6. Ponemon, P.L.: Security of cloud computing users. Ponemon institute, research report. http://www.ca.com/files/industryresearch/security-cloud-computing-users_235659.pdf (May 2010)
7. Gens, F.: New IDC IT cloud services survey: top benefits and challenges. IDC exchange. http://blogs.idc.com/ie/?p=730 (2009)
8. Almulla, S.A., Yeun, C.Y.: Cloud computing security management. II International Conference on Engineering Systems Management and its Applications (ICESMA) (2010)

9. Celesti, A., Tusa, F., Villari, M., Puliafito, A.: Security and cloud computing: inter cloud identity management infrastructure. 19th IEEE International workshop on Enabling Technologies: Infrastructures for Collaborative Enterprises (WETICE), 2010, pp. 263–265

10. Rui, J.: Advanced secure authentication framework for cloud computing. Int. J. Smart Sens. Intell. Syst. 6(4), (2013)

11. Jeremy, K.: One of the most convincing phishing attacks yet tricks you with Dropbox sharing. PCWorld. http://www.pcworld.com/article/2835892/dropbox-used-for-convincing-phishing-attack.html (Oct, 20 2014)

12. Robert, M., Oops! amazon web services customer unleashes 'denial of money' attack—on himself. WIRED. http://www.wired.com/2012/04/aws-bill-in-minutes/ (April, 2012)

13. CRM Provider Salesforce Hit With Malware Attack: Entrust. http://www.entrust.com/crm-provider-salesforce-hit-malware-attack/ (September, 2014)

14. Darreni, P.: Google app engine has thirty flaws, says researcher. The register. http://www.theregister.co.uk/2014/12/09/google_app_engine_has_thirty_flaws_says_researcher/ (December 2014)

15. Jiang, R.: Advanced secure user authentication framework for cloud computing. Int. J. Smart Sens. Intell. Syst. 6(4), (2013)

16. Nayak, S.K., Mohapatra, S., Majhi, B.: An improved mutual authentication framework for cloud computing. IJCA 52(5), (2012)

17. Takabi, H., Joshi, J.B.D., Ahn, G.: SecureCloud: towards a comprehensive security framework for cloud computing environments. Proceedings of IEEE 34th Annual Computer Software and Application Conference Workshops, pp. 393–398, 19–23 July 2010

18. Falas, T., Kashani, H.: Two-dimensional bar-code decoding with camera-equipped mobile phones. Proceedings of the Fifth Annual IEEE International Conference on Pervasive Computing and Communications Workshops, pp. 597–600, 19–23 March, 2007

19. OASIS: Security assertion mark up language, V2.0, Technical overview. http://docs.Oasis-open.org/Security/Saml/Post2.0/sstc-saml-tech-overview-2.0-cd-02.html

KPS: A Fermat Point Based Energy Efficient Data Aggregating Routing Protocol for Multi-sink Wireless Sensor Networks

Kaushik Ghosh, Pradip K. Das and Sarmistha Neogy

Abstract Lifetime of a multi-sink wireless sensor network (WSN) may increase considerably when data aggregation is introduced in a Fermat point based routing protocol. However, data aggregation should come with a cost of delay in packet forwarding time. It has been seen that increasing the transmission distance could increase the network lifetime. However, our observation shows that after a certain point, lifetime readings of the network would dip for a distance vector type of protocol. Thus, it becomes necessary to choose an appropriate aggregation factor and transmission radius depending upon the requirement of the application for which the WSN has been installed. In this paper we have presented a Fermat point based data aggregating protocol which is distance vector protocol in nature. We have compared its lifetime with some other Fermat point based protocols and studied the effect of aggregation factor on cumulative delay for packet forwarding. Moreover, effect of increased transmission radius on the lifetime of the proposed protocol too was studied.

Keywords Multi-sink WSN · Aggregation factor · Network lifetime · Fermat point based routing protocols · Cumulative delay

K. Ghosh (✉)
Mody University of Science and Technology, Rajasthan, India
e-mail: kaushikghosh.fet@modyuniversity.ac.in

P.K. Das
RCCIIT, Kolkata, India
e-mail: pkdas@ieee.org

S. Neogy
Jadavpur University, Kolkata, India
e-mail: sneogy@gmail.com

© Springer India 2016
R. Chaki et al. (eds.), *Advanced Computing and Systems for Security*,
Advances in Intelligent Systems and Computing 395,
DOI 10.1007/978-81-322-2650-5_13

1 Introduction

Due to the inherent hardware constraint of the sensor nodes, energy efficiency is the most talked about area for wireless sensor networks (WSN). A WSN comprises of large number of sensor nodes placed over a sensor field whose number may be several orders of magnitude higher than that of traditional ad hoc networks. The data transferred by the sensor nodes are collected through gateway nodes in the sink(s). It is at the sink(s) where the data gathered by the nodes are analyzed and appropriate action is taken thereafter. The sinks are generally proper nodes with adequate computation and communication facility, memory space and connected with a power socket. Sensor nodes (along with gateway nodes) on the other hand, however, differ from the sinks in computation, communication, primary memory size, and nature of the power source.

Each sensor node comprises of four basic units—(i) a sensing unit, (ii) a processing unit, (iii) a transceiver unit, and (iv) a power unit—in a package of around one cubic centimeter [1]. This size limitation of the sensor nodes put constraints on its computing power, energy expenditure, and memory units. Of these three units again, the constraint on power is the greatest constraint. Because for certain applications like military surveillance and environment monitoring [2], hazardous deployment environment have limited/eliminated the scope for battery replacement. These factors have thus motivated researchers to come out with means of prolonging network lifetime to waive out the cost of network deployment for a particular application.

Although most of the energy efficient routing protocols for WSNs are designed keeping a single sink in mind only, yet a number of protocols for multiple sinks too are coming up. A major drawback of a single sink WSN is single point failure. If the sink goes down for some reason, the entire exercise of data collection and transfer by the sensors becomes fruitless. Moreover, there are certain applications where presence of multiple sinks is a matter of necessity and not of redundancy. Other than single point failure, following factors too lead to place multiple sinks in a WSN as a replacement of single sink:

(i)   issues in data acquisition with a single sink
(ii)   load balancing
(iii)   reduction in sink hole problem
(iv)   reduction in total number of hops encountered by a packet
(v)   providing infrastructure support over multiple interfaces and
(vi)   increase in overall network lifetime

Of the different application areas of WSNs, precision agriculture has been the one among the applications in the forefront. It demands intensive field data acquisition where the nodes in WSN report their measurements to a collector point [3]. In some of the cases the collector point is the sink. In many other cases, a sensor node is used for data aggregation and the aggregated data is then transmitted to the sinks [4]. In both the cases, however, the presence of multiple sinks may turn out to be a necessity.

To reduce the energy consumption in a WSN, it is first required to know the major factor that consumes the major chunk of energy there. Different works [5, 6] confirm that all of the factors, transmitting distance is the one which affects network lifetime more than any other factors present. This is due to the fact that with increase in transmitting distance, energy consumption increases superlinearly [7–10].

A Fermat point based routing protocol fits into address the presence of multiple sinks [11] along with reducing the energy consumption in a WSN by reducing overall transmitting distance [12]. Other than transmission distance, volume of data transferred too plays an important role in the lifetime of a WSN—it is directly proportional to the amount of energy consumed [6]. Data aggregation is a technique which can reduce energy consumption in a WSN by controlling the total number of transmissions [13]. Combining the concepts of both data aggregation and Fermat point can thus enhance network lifetime to a considerable extent by attacking these two major components of energy consumption [11, 14, 15].

In this paper we have introduced a Fermat point based data aggregating protocol —the KPS protocol, named by taking the initials of the authors. Here we have tried to understand the effect of cumulative delay in the network due to enhanced degree of aggregation along with effect of increased transmission radius on the lifetime of a multi-sink WSN.

The rest of the paper has been organized as follows: Sect. 2 contains related works. In Sect. 3 we introduce our protocol. Section 4 contains results and finally we conclude in Sect. 5.

2 Related Works

As mentioned in the previous section, a Fermat point based data aggregating routing protocol would surely reduce energy consumption and thereby enhance lifetime of a WSN. The Fermat point is a point within a triangle or polygon, such that the sum of the distances of the point from all the vertices of the triangle or polygon is minimum, when compared to the sum of any other point within the boundary of that triangle/polygon to all the vertices. Motivated by this, researchers have designed Fermat point based routing protocols for multiple target regions/sinks to reduce the overall transmission distance [7, 8, 11, 12, 16]. The node closest to the theoretical Fermat point has been marked as the Fermat node (FN) and all transmissions to the sinks have been made through FN only [11].

Data aggregating schemes too has helped in energy conservation in WSNs [13]. Here, the authors have proposed a data centric approach of routing to remove the energy constraints of WSNs. The impact of source–destination placement and network density on the energy costs and delay associated with data aggregation was examined in [17].

Contrary to the traditional approach, [18] proposes multiple initiators for each data-gathering iteration and claims that their proposed approach will consume energy in a uniform way when compared with serial data aggregation schemes [19, 20].

In [21] the authors present an exact as well as approximate algorithm to find the minimum number of aggregation points for maximizing the network lifetime.

He et al. [22] proposed an adaptive application-independent data aggregation scheme. The aggregation decisions are kept in a module between network and data link layer. The protocol reduces end-to-end delay by 80 % and transmission energy consumption by 50 %. Privacy preserving data aggregation schemes for additive aggregation functions was proposed in [23] to bridge the gap between collaborative data collection by WSN and data privacy.

Combining therefore the concepts of both Fermat point and data aggregation is going to reduce overall network energy consumption till a great extent. It is very natural to think of selecting the Fermat point as the aggregation point [24]. Although certain schemes have followed a tree-based approach [12], authors in [24] showed that it does not require any tree to be constructed for aggregation. Son et al. [25] too is a good example of a Fermat point based data aggregation scheme for data forwarding. In both [24, 25] the authors have compared their results with a greedy incremental tree (GIT) forwarding. In both the cases the Fermat point based scheme recorded higher lifetime as compared to the tree-based scheme.

Authors in [26–31] have discussed about placing multiple sinks in a WSN and their effect on network lifetime. Some of the algorithms here bank upon optimum sink placement for increasing network lifetime. Although optimal sink placement is a special case and cannot be proposed for any application whatsoever, yet it is worth discussing their effects on lifetime of multi-sink WSNs.

Kim et al. [26] finds the optimal position of multiple sink nodes and the optimal traffic flow from the different sensors to the sinks. The scheme proposed outperforms the multi-sink aware minimum depth tree (m-MDT) scheme in terms of network lifetime. Authors in [27] have shown that with lapse of time (i) the number of disconnected regions in a network increases and (ii) the failure of nodes close to the sink also increases. They prove that reconstruction of the minimum energy tree after the occurrence of energy failures can prolong network lifetime. The results in the paper show a decrease in the percentage of exhausted nodes for an increase in the number of sink nodes. Flathagen et al. [28] shows that in comparison between topology aware and geo-aware placement strategies for sinks, the topology-aware placement outperforms the geo-aware placement with lifetime as the performance parameter. The use of routing metric placement (RMP) results in lifetime increases of 60 % for two sinks and 25 % for three sinks compared to the use of K-means placement based on the K-means clustering algorithm. Carels et al. [29] proposes employing a virtual root for coordination between multiple sinks. This paper shows that the concept of a virtual root can work and can be implemented with a minimal complexity. Authors here have shown that for an increase in the number of sinks from 1 to 4, maximum energy decreased by 45 % and average energy decreased by more than 30 %. For the real-life tests, the average energy consumption decreased with more than 30 % and with more than 50 % for the maximal energy consumption when the number of sinks was increased from 1 to 2. [30] uses a mathematical model to determine the optimal position of the sink nodes in a WSN and thus minimizes the sensors' average distance to the nearest sink. The paper presented two positioning algorithms: global algorithm and 1hop algorithm. In the global algorithm, the information of all the nodes in the network is used and in the 1hop algorithm only information of the neighboring nodes is used. To balance the traffic load and to prevent energy depletion of nodes close to the sink, the paper presents the 1hop relocation algorithm. Through simulations it was shown that the network lifetime may be doubled as compared to the static or random deployment. A genetic algorithm is used in [31] for sensor-sink binding in multi-sink WSNs. The work kept overloading of sinks in mind for cases where sensors tend to bind to a particular sink to reduce communication distance. Through a genetic algorithm-based approach authors have solved the problem of balancing the load of sensors amongst sinks in a multi-sink WSN, while ensuring that the best routes to sinks are found for the sensors that cannot directly reach a sink.

3 Proposed Protocol

In this paper we propose a Fermat point based data aggregating routing protocol. We name the protocol as KPS, taking the initials of the authors. In this protocol, the nodes are either deployed in a grid fashion (see Fig. 1) or in a random fashion. After deployment, a node is supposed to find out the theoretical Fermat point for the logical polygon/triangle formed by sender node and the sinks. For this paper, we have considered the sinks to be located at three vertices of a rectangular sensor field. After the theoretical Fermat point has been found, the node closest to theoretical Fermat point is marked as FN by the source node. Figure 2, redrawn from our previous work [11], depicts the same. The algorithm for finding the Fermat point is the Minima algorithm, proposed in one of our earlier works [7]. Figure 3, redrawn from [7] gives the steps of Minima algorithm. The Minima algorithm was proposed for finding the Fermat point for multiple geocast regions, but may be extended for multiple sinks/target regions. Table 1 gives the description of variables used in the algorithm and Table 2 describes the functions used there. Both the tables are taken from [7].

After a source comes to know about its immediate neighbors and the FN, the transmission phase may start. Transmission from a source to a sink would go on in multiple hops. The transmission phase may again be divided into two parts: (i) transmission from sender to FN and (ii) from FN to sinks. In either phase, before a node can forward packet to one of its 1-hop neighbors, it calculates the forwarding potential κ, of all its 1-hop neighbors [11]. The forwarding potential of a neighbor is calculated as

$$\kappa = \text{res\_energy}/\text{dist} \qquad (1)$$

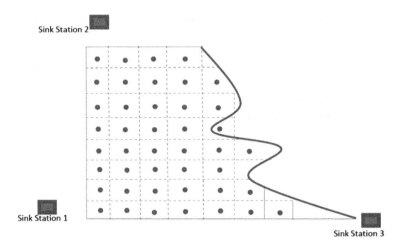

Fig. 1 A sensor field with sinks placed at its extreme corners

Fig. 2 Source node, Fermat
node, and sinks

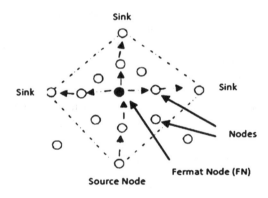

where,

res_energy = residual battery power of a node in millijoules.
dist = distance of a node from a particular sink.

The neighbor with highest value of κ is selected for forwarding a packet. Since after every transmission/forwarding, residual energy of a node is going to change, the value of κ thus has to be calculated for neighbors for every transmission/ forwarding.

This is important in WSNs, where the topology remains fixed due to immobility of nodes. If a node keeps choosing the same neighbor over and over again for packet forwarding, energy depletion would be fast and lifetime of a network would thus reduce. For neighbor discovery, a node will transmit **neighbor_beacon** control packets. A neighbor_beacon packet would contain node ID of source and coordinates of its FN. In reply to those neighbor_beacons, the neighbors would reply with

Minima Algorithm

Input: Coordinates of the sender node and that of different geocast regions.
Output: (fx, fy). Coordinates of the Fermat_Point.
Tdist : Total distance traveled by the packet.
Tpow: The sum of power consumed by all the intermediate nodes to
forward the packet to m geocast regions.

```
1.    max_x=MAX_x (Sx, GRX(N))
2.    max_y=MAX_y (Sy, GRY(N))
3.    min_x=MIN_x (Sx, GRX(N))
4.    min_y=MIN_y (Sx, GRY(N))
5.    dx=0;   /*          Initialize dx */
6.    dy=0    /* Initialize dy */
7.    flag=0; /* To check the Fermat Point*/
8.    for ( i=min_x; k<max_x; i++ )
9.    { if (flag==1) break; /* Fermat point found */
10.   for ( j=min_y; j=max_y; j++ )
11.   { x=i;
12.   y=j;
13.   for ( k=0;k<n;k++ )
14.   { dx+= termdx ( x,GRX(k),y,GRY(k) );
15.   dy+= termdy ( y,GRY(k),x,GRX(k) );
16.   } /* end of for loop (line-11)*/
17.   if ( dx==0 && dy==0 )
18.   { flag=1; /* Fermat point found */
19.   break;
20.   }
21.   dX=0 dY=0;
22.   } /* end of for loop (line-8) */
23.   }/* end of for loop (line – 6) */
24.   if( flag==1 )
25.   { fx=x; fy=y; } /* Fermat point */
26.   Tdist= Total_Dist ( Sx, Sy, GRX(N), GRY(N), fx, fy );
27.   Tpow= Total_Pow ( Tdist )
```

Fig. 3 Minima algorithm for finding Fermat point of a triangular/polygonal region

their node ID, distance from the FN and their individual residual energies. This way κ for all the neighbors would be calculated by a source node.

In the second phase of transmission, the FN would take one sink at a time for transmission and would calculate forwarding potential in the same way as discussed for the previous step. Before FN could start forwarding data packets, it would have to look for the value of aggregation factor (AGFACT) decided upon.

The value of AGFACT determines the number of packets FN is supposed to aggregate since we have considered FN as the aggregation point. In fact more the value of AGFACT, more is the reduction in transmission and thereby more would be the lifetime of the network concerned. However, a trade-off between lifetime enhanced and the incurred delay due to aggregation has to be made.

Another matter of concern with a Fermat point/Fermat node-based routing protocol is the extra burden of computation and communication incurred on a node while it acts as Fermat node for one or multiple source node(s). This extra burden would thus lead to deplete energy in nodes acting as Fermat nodes at a much faster

Table 1 Variables used in Minima algorithm

Symbols	Definitions
max_x	Maximum value of X coordinates
max_y	Maximum value of Y coordinates
min_x	Minimum value of X coordinates
min_y	Maximum value of Y coordinates
(Sx, Sy)	Coordinates of the sender
n	Number of geocast regions
GRX [ ]	Array storing the values of X-coordinates of all the geocast regions
GRY [ ]	Array storing the values of Y-coordinates of all the geocast regions
dx	Minimum value of X
dy	Minimum value of Y
(fx, fy)	Coordinates of the fermat point
Tdist	Total distance of transmission
Tpow	Total power consumed to transmit a packet from the sender to all the geocast regions

rate as compared to other nodes. As a result, lifetime of a network would get affected. As a remedy to this problem we introduce a new parameter δ, which keeps the measure of percentage of energy depleted in a node acting as Fermat node. When the value of δ reaches 100 % for a node, it is understood to be depleted out of its entire energy. And according to the definition of network lifetime selected by us, that is, the point of time when we declare the network to be dead.

We thus recommend to relieve a Fermat node of its extra duties when its residual energy comes below a certain threshold fraction/percentage, τ, in order to maximize network lifetime. When δ equals to τ, we propose to choose a second Fermat node—a node that is second closest to the theoretical Fermat point—for data aggregation and

Table 2 Functions used in Minima algorithm

Functions	Definitions
MAX_x(Sx,GRX[])	**Input:** X coordinate of the sender and X coordinates of all the geocast regions. **Output:** Maximum value amongst the X coordinates.
MAX_y (Sy,GRY[])	**Input:** Y coordinate of the sender and Y coordinates of all the geocast regions. **Output:** Maximum value amongst the Y coordinates.
MIN_x(Sx,GRX[])	**Input:** X coordinate of the sender and X coordinates of all the geocast regions. **Output:** Minimum value amongst the X coordinates.
MIN_y(Sy,GRY[])	**Input:** Y coordinate of the sender and Y coordinates of all the geocast regions. **Output:**Minimum value amongst the Y coordinates.
termdx(x,GRX[], y,GRY[])	$$\frac{\delta}{\delta x}f(x)\|y=constant$$ **Input:** X and Y coordinate of any point within the polygon and (X,Y) coordinates of all the geocast regions. **Output:** dx i.e., $$\frac{\delta}{\delta x}f(x)\|y=constant$$
termdy(y,GRY[], x,GRX[])	$$\frac{\delta}{\delta y}f(y)\|x=constant$$ **Input:** X and Y coordinate of any point within the polygon and (X,Y) coordinates of all the geocast regions. **Output:** dy i.e., $$\frac{\delta}{\delta y}f(y)\|x=constant$$
Total_Dist(Sx, Sy, GRX[], GRY[], fx, fy)	**Input:** (X,Y) Coordinates of the sender, (X,Y) coordinates of all the geocast regions and (X,Y) coordinates of the Fermat point. **Output:**Total distance from the sender to all the geocast regions via the Fermat point, taken individually.
Total_Pow(Tdist)	**Input:**Total distance of the path from sender to the Fermat point and from there to all the geocast regions taken individually. **Output:** Total power consumed for transmitting a packet to all the geocast regions.

forwarding. The second Fermat node may again be deselected for a third Fermat node if its percentage of depleted residual energy, δ, falls below the prescribed value of τ.

Radio Model The radio model discussed here is same as in [11]. The different sources of energy consumption considered here are: (i) sensing (E_{Sensing}), (ii) computation ($E_{\text{computation}}$), (iii) forwarding ($E_{\text{forwarding}}$), (iv) receiving ($E_{\text{receiving}}$), and (v) listening ($E_{\text{listening}}$). Energy consumed by the source node for transmission (E_{TX}) requires a node to sense the physical parameter (E_{Sensing}), do necessary computation for converting the physical parameter into transmissible form ($E_{\text{computation}}$), and then forward it to the next hop forwarder ($E_{\text{forwarding}}$). Relay nodes on the other hand consumes energy to receive a packet ($E_{\text{receiving}}$) and then forward it further ($E_{\text{forwarding}}$). Again, when a node neither acts as a sender nor is relaying any packet for any sender, then too it is consuming energy while listening to the transmissions of its neighbors ($E_{\text{listening}}$).

A node is considered to be in on state, while it acts as either sender or relay and in off state while listening to transmission of other nodes. The "on" period of a node (t_{on}) is thus defined as the time it is engaged in either transmitting (T_{tx}) or forwarding (T_{fwd}) data. The time for which a node is in listening mode (T_{lst}) is considered as its "off" period (t_{off}). The duty cycle D of a node is given as

$$D = t_{\text{on}}/(t_{\text{on}} + t_{\text{off}}) \tag{2}$$

Now, more the number of 1-hop neighbors, greater are the chance of KPS selecting different neighbors for subsequent transmissions (see Eq. 1). Selecting different neighbors for different transmission/forwarding one way is helpful in increasing the network lifetime—no single node is overtaxed for data forwarding.

In [30] the authors have shown how lifetime of a WSN may increase with increase in transmission range (TXR). It happens due to reason stated above. But increasing number of neighbors that come due to increase in TXR would have an adverse effect too on the network lifetime—a node has to transmit/forward data over a greater distance. Considering the value of path loss exponent as 3, increase in transmitting distance beyond a certain point surely is going to take a toll on lifetime of the network. This observation could be seen in the next section.

From [32], we have taken the value of $E_{\text{computation}}$ as 117 nJ/bit. Similarly, E_{sensing} is taken as 1.7 µJ/bit [33]. $E_{\text{listening}}$ on the other hand is not a function of number of bits transmitted. Rather, it depends upon the number of seconds spent in listening mode and its value is taken as 570 µJ/s [34].

The radio model thus can finally be expressed as

$$E_{\text{TX}} = m * 117 * 10^{-9} + m * 1.7 * 10^{-6} + D * m * \varepsilon * d^n \tag{3}$$

$$E_{\text{forwarding}} = D(m * E + m * \varepsilon * d^n) \tag{4}$$

$$E_{\text{listening}} = (1 - D) * 570 * 10^{-6} \tag{5}$$

where

m	Packet size in number of bits
n	Path loss exponent
D	Duty cycle
E	50 nJ/bit
E	8.854 pJ/bit/m2
D	Internodal distance

The values of E and ε are taken from [6]. ε stands for permittivity, and E is the minimum start-up energy required for any communication.

4 Results

The simulation has been done in C (gcc compiler). Assuming, one transmission per hour, we present the performance of our proposed protocol compared to some other Fermat point based protocols, in this section.

As discussed in [11], deployment pattern, source selection mode, and aggregation scheme play a considerable effect on the lifetime of a WSN. Lifetime of a network has been considered as defined in . It is the time recorded from network deployment till the time the first node goes out of energy. As reference, we take Fig. 4a, b from our previous work [11], which shows the performance of our protocol with Fermat point based variants of Greedy forwarding (F-Greedy), Compass routing (F-Compass), and Residual energy-based forwarding (F-residual). In a nutshell, F-Greedy, F-Compass, and F-Residual follows the basic postulates of the parent algorithms and add on the concept of Fermat point based data forwarding. Just like KPS, F-Greedy, F-Compass, and F-Residual too transmit data to sinks in two phases: first, to Fermat node and then to the sinks. For all four protocols, we considered the nodes to be deployed in a random fashion. Table 3 gives the parameters used for readings in Fig. 4a, b. As an extension of work for results obtained in Fig. 4a, b, we have compared our protocol with the same protocols for transmission ranges 90 and 100 m in Fig. 4c, d respectively.

Figure 4c, d shows the result for the same. Network parameters for Fig. 4c, d (other than transmission range (**TXR**)) are same as given in Table 3. We can see that F-greedy and F-residual protocols have record comparable lifetimes for TXR 80, 90, and 100 m and have been outperformed by both KPS and F-compass routing. Narrowing down our discussion to KPS and F-compass routing, we see that for TXR value of 80 m, lifetime recorded for KPS protocol was higher as compared to F-compass routing. At TXR = 90 m, F-compass routing started showing better results at higher degrees of aggregation and at TXR = 100 m, it completely outperformed KPS.

The reason behind this appears to be due to the fact that, KPS is basically a distance vector routing—lesser the distance of a node from a particular sink, greater is the chance of getting selected as the forwarder. Therefore, with increase in TXR, a node has more and more option of neighbors for data forwarding and this

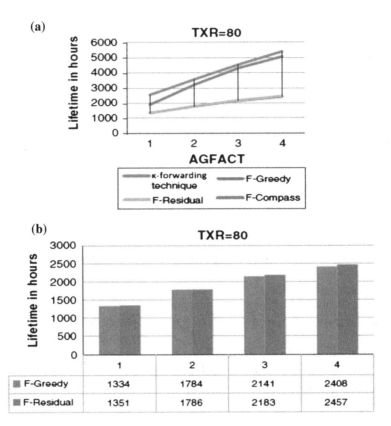

Fig. 4 a, b Lifetime comparison between different forwarding techniques (TXR = 80 m). **c, d** Performance of KPS protocol compared to other Fermat point based protocols (TXR = 90 and 100 m)

increases the possibility of selecting a "far away" neighbor as compared to neighbors residing nearby. According to us, having choice of a greater number of neighbors due to increase in TXR would have twofold effect on the lifetime of a WSN. Now, more the number of 1-hop neighbors, greater are the chance of KPS selecting different neighbors for subsequent transmissions (see Eq. 1). Selecting different neighbors for different transmission/forwarding one way is helpful in increasing the network lifetime—no single node is overtaxed for data forwarding.

In [36] the authors have shown how lifetime of a WSN may increase with increase in TXR. It happens due to reason stated above. But increasing number of neighbors that come due to increase in TXR would have an adverse effect too on the network lifetime—a node has to transmit/forward data over a greater distance. With the value of path loss exponent as 3, increase in transmitting distance beyond a certain point surely is going to take a toll on lifetime of the network.

For KPS with network parameters as in Table 3, we could see that 100 m is that point where the advantage of traffic distribution among different neighbors is

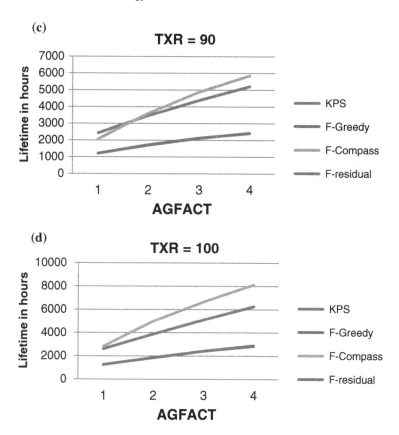

Fig. 4 (continued)

outplayed by energy consumption due to transmission over a greater distance, in case the nodes are deployed randomly. However, for grid deployment, KPS outperforms F-Compass Routing at TXR value as high as 100 m (Fig. 5). As it had been seen in [11] that lifetime of a WSN would vary for random and grid deployment, in Fig. 5 we show the performance of the discussed protocols for grid deployment as well. The network parameters changed for results of Fig. 5 are as Table 4, rest of the parameters is same as in Table 3. According to our observation, random deployment of nodes cannot give a conclusive report on the performance of a protocol when lifetime has been taken as the evaluating parameter.

However, grid deployment and round robin source selection being too ideal an assumption for many applications, it is to be noted that random node deployment and random source selection may be considered for most of the practical cases.

Figure 6 shows how network lifetime may be increased by judiciously selecting the threshold point at which a Fermat node may be relieved of its duties. For results of Fig. 6, we have considered random node deployment and random source selection mechanism. The value for AGFACT was kept as 2. All other parameters

Table 3 Different network parameters for random node deployment

Parameters	Values
Nodes	100
Area	150 m × 150 m
Number of sinks	3
Deployment pattern	Random
Transmission range (TXR)	80 m
Data rate	38.4 kbps
Packet size (p)	36 bytes
Initial energy of nodes	1 J
Source selection mode	Random
Aggregation scheme	Converts n input packets to one packet of size $n \times p$

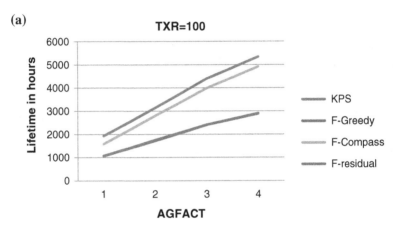

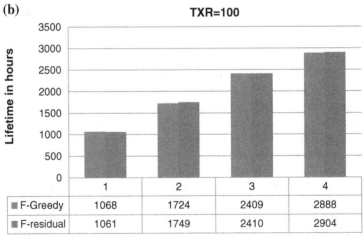

Fig. 5 **a** and **b** Lifetime comparison for grid deployment

New/changed parameters	Values
Deployment pattern	Grid
Individual grid size	30 × 15
Number of nodes	50
Transmission range (TXR)	100 m

Table 4 New/changed parameters changed for grid deployment

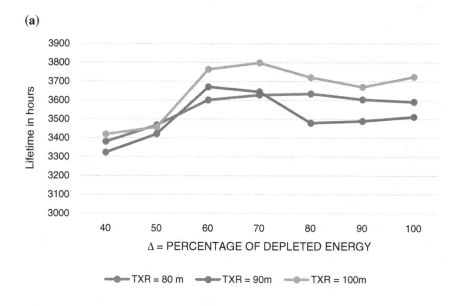

(a)

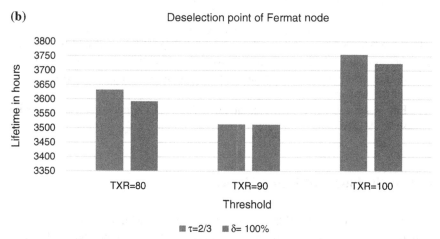

(b)

Fig. 6 **a** Lifetime recorded for different values of δ. **b** Lifetime comparison with τ at two-third of the depleted energy

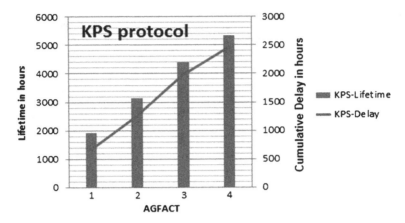

Fig. 7 Lifetime and cumulative delay for KPS protocol

remain same as before. Figure 6a shows that choosing τ as 80 % maximizes network lifetime of a multi-sink WSN when TXR = 80 m. Whereas, when TXR is 90 m, maximum network lifetime is recorded when we deselect a Fermat node at its 60 % energy depletion. Again for TXR = 100 %, maximum lifetime was recorded at τ equals to 70 %. Thus, as a rule of thumb, we consider shedding off the responsibility of a Fermat node when two-third of its energy gets depleted. Figure 6b ensures us that when we deselect a Fermat node at τ equals to 2/3, we record a higher lifetime as compared to the scenario where we let the Fermat node to perform its duties till it depletes its 100 % energy reserve.

Another obvious observation for all the protocols is—lifetime is bound to increase with increased degree of aggregation. But as the value of AGFACT would keep on increas*ing, the aggregating node (FN in our case) would have to wait more for all the packets* to arrive, before it can aggregate and forward them to different sinks. Moreover, with increase in the value of AGFACT, more would be the time consumed for computation. This in turn would increase the cumulative delay in the network. Cumulative delay is the summation of total delay experienced by all the nodes in the network.

In Fig. 7 we can see how the cumulative delay of the network increases with increase in AGFACT.

5 Conclusions and Future Work

Lifetime of a multi-sink WSN is bound to increase with increased degree of aggregation. The result would be even more impressive if we could reduce the total transmission distance by using a Fermat point based routing protocol on top of data aggregation. However, improvement in network lifetime reflected due to aggregation would come at the cost of increased cumulative delay in the network. It is thus important to understand the network requirement before selecting a value for AGFACT. For delay sensitive networks, it is advisable to select value for AGFACT as low as possible.

One important observation in this paper is—increased number of neighbors due to increase in transmission radius may come with a cost along with its obvious advantage. More number of neighbors is good in a way as a node would have more options to choose for the next hop forwarder. Moreover, due to the parameters in forwarding potential in KPS, a node would end up choosing a different forwarder for subsequent transmissions or forwarding. This way, disadvantages of protocols following greedy forwarding are eliminated, where the same neighbor is used over and over again for packet forwarding. But the flip side is—selecting a neighbor far away is going to deplete residual energy of a node to such a degree that the advantage discussed earlier may get nullified. Thus, selecting appropriate transmission radius for a node is both required and important for network lifetime enhancement.

Another finding of importance is—random deployment of nodes cannot remark conclusively on the performance of a protocol when lifetime is the evaluating parameter. With change in transmission distance and aggregation factor, comparative performance graphs for lifetime between two or more protocols would change positions and thereby it becomes impossible to give any conclusive statement. Grid deployment on the other hand gives consistent results for lifetime between different protocols. However, for a major set of applications in WSNs both grid deployment and round robin source selection may not be either feasible or desirable. Thus, random deployment and random source selection may not be ignored for many practical purposes.

As future work we would carry on some more comparison of the proposed work with protocols involving optimum sink placement for lifetime enhancement. A detailed experimentation on selecting a proper value of τ for lifetime maximization has to be carried out as well. Moreover, an optimum transmission range for the proposed protocol has to be found out, for which it would generate maximum network lifetime.

References

1. Pottie, G.J., Kaiser, W.J.: Wireless integrated network sensors. Commun. ACM **43**(5), 51–58 (2000)
2. Arampatzis, T., Lygeros, J., Manesis, S.: A survey of applications of wireless sensors and wireless sensor networks. In: Intelligent Control, 2005. Proceedings of the 2005 IEEE International Symposium on, Mediterranean Conference on Control and Automation, pp. 719–724. IEEE (2005)
3. Camilli, A., Cugnasca, C.E., Saraiva, A.M., Hirakawa, A.R., Corrêa, P.L.P.: From wireless sensors to field mapping: anatomy of an application for precision agriculture. Comput. Electron. Agric. **58**(1), 25–36 (2007)
4. Wark, T., Corke, P., Sikka, P., Klingbeil, L., Guo, Y., Crossman, C., Valencia, P., Swain, D., Bishop-Hurley, G.: Transforming agriculture through pervasive wireless sensor networks. IEEE Pervasive Comput. **6**(2), 50–57 (2007)

5. Gandham, S.R., Dawande, M., Prakash, R., Venkatesan, S.: Energy efficient schemes for wireless sensor networks with multiple mobile base stations. In: IEEE Global Telecommunications Conference, 2003, GLOBECOM'03, vol. 1, pp. 377–381. IEEE (2003)
6. Heinzelman, W.B., Chandrakasan, A.P., Balakrishnan, H.: An application-specific protocol architecture for wireless microsensor networks. IEEE Trans. Wireless Commun. **1**(4), 660–670 (2002)
7. Ghosh, K., Roy, S., Das, P.K.: An alternative approach to find the fermat point of a polygonal geographic region for energy efficient geocast routing protocols: global minima scheme." In: NETCOM'09. First International Conference on Networks and Communications, 2009, pp. 332–337. IEEE (2009)
8. Lee, S.-H., Ko, Y.-B.: Geometry-driven scheme for geocast routing in mobile ad hoc networks. In: IEEE 63rd Vehicular Technology Conference, 2006. VTC 2006-Spring, vol. 2, pp. 638–642. IEEE (2006)
9. I-Shyan, H., Pang, W.-H.: Energy efficient clustering technique for multicast routing protocol in wireless adhoc networks. IJCSNS **7**(8), 74–81 (2007)
10. Alam, N., Balaie, A.T., Dempster, A.G.: Dynamic path loss exponent and distance estimation in a vehicular network using doppler effect and received signal strength. In: 2010 IEEE 72nd Vehicular Technology Conference Fall (VTC 2010-Fall), pp. 1–5. IEEE (2010)
11. Ghosh, K., Das, P.K., Neogy, S.: Effect of source selection, deployment pattern, and data forwarding technique on the lifetime of data aggregating multi-sink wireless sensor network. In: Applied Computation and Security Systems, pp. 137–152. Springer, India (2015)
12. Ssu, K.-F., Yang, C.-H., Chou, C.-H., Yang, A.-K.: Improving routing distance for geographic multicast with Fermat points in mobile ad hoc networks. Comput. Netw. **53**(15), 2663–2673 (2009)
13. Krishnamachari, L., Estrin, D., Wicker, S.: The impact of data aggregation in wireless sensor networks. In: Proceedings of the 22nd International Conference on Distributed Computing Systems Workshops, pp. 575–578. IEEE (2002)
14. Son, J., Pak, J., Han, K.: Determination of aggregation point using Fermat's point in wireless sensor networks. APWeb Workshops 2006, LNCS, vol. 3842, pp. 257–261 (2006)
15. Son, J., Pak, J., Kim, H., Han, K.: A decentralized hierarchical aggregation scheme using Fermat points in wireless sensor networks. Evo Workshops 2007, LNCS, vol. 4448, pp. 153–160 (2007)
16. Song, Y.-M., Lee, S.-H., Ko, Y.-B.: Ferma: an efficient geocasting protocol for wireless sensor networks with multiple target regions." In: Embedded and Ubiquitous Computing–EUC 2005 Workshops, pp. 1138–1147. Springer, Berlin (2005)
17. Intanagonwiwat, C., Estrin, D., Govindan, R., Heidemann, J.: Impact of network density on data aggregation in wireless sensor networks. In: Proceedings of 22nd International Conference on Distributed Computing Systems, 2002, pp. 457–458. IEEE (2002)
18. Massad, Y.E., Goyeneche, M., Astrain, J.J., Villadangos, J.: Data aggregation in wireless sensor networks. In: 3rd International Conference on Information and Communication Technologies: From Theory to Applications, ICTTA 2008, pp. 1–6. IEEE (2008)
19. Patil, S., Das, S.R., Nasipuri, A.: Serial data fusion using space-filling curves in wireless sensor networks. In: 2004 First Annual IEEE Communications Society Conference on Sensor and Ad Hoc Communications and Networks, 2004. IEEE SECON 2004, pp. 182–190. IEEE (2004)
20. Dai, X., Xia, F., Wang, Z., Sun, Y.: A survey of intelligent information processing in wireless sensor network. In: Mobile Ad-hoc and Sensor Networks, pp. 123–132. Springer Berlin (2005)
21. Al-Karaki, J.N., Ul-Mustafa, R., Kamal, A.E.: Data aggregation in wireless sensor networks—exact and approximate algorithms. In: Workshop on High Performance Switching and Routing, 2004, HPSR, pp. 241–245. IEEE (2004)
22. He, T., Blum, B.M., Stankovic, J.A., Abdelzaher, T.: AIDA: adaptive application independent data aggregation in wireless sensor networks. ACM Trans. Embed. Comput. Syst. (TECS) **3**(2), 426–457 (2004)

23. He, W., Liu, X., Nguyen, H., Nahrstedt, K., Abdelzaher, T.T.: PDA: privacy-preserving data aggregation in wireless sensor networks. In: 26th IEEE International Conference on Computer Communications, INFOCOM 2007, pp. 2045–2053. IEEE (2007)
24. Son, J., Pak, J., Han, K.: Determination of aggregation point using Fermat's point in wireless sensor networks. APWeb Workshops 2006, LNCS, vol. 3842, pp. 257–261 (2006)
25. Son, J., Pak, J., Kim, H., Han, K.: A decentralized hierarchical aggregation scheme using fermat points in wireless sensor networks. In: Applications of Evolutionary Computing, pp. 153–160. Springer, Berlin (2007)
26. Kim, H., Seok, Y., Choi, N., Choi, Y., Kwon, T.: Optimal multi-sink positioning and energy-efficient routing in wireless sensor networks." In: Information Networking. Convergence in Broadband and Mobile Networking, pp. 264–274. Springer, Berlin (2005)
27. Oyman, E.I., Ersoy, C.: Multiple sink network design problem in large scale wireless sensor networks. In: 2004 IEEE International Conference on Communications, vol. 6, pp. 3663–3667. IEEE (2004)
28. Flathagen, J., Kure, Ø., Engelstad, P.E.: Constrained-based multiple sink placement for wireless sensor networks. In: 2011 IEEE 8th International Conference on Mobile Adhoc and Sensor Systems (MASS), pp. 783–788. IEEE (2011)
29. Carels, D., Derdaele, N., Poorter, E.D., Vandenberghe, W., Moerman, I., Demeester, P.: Support of multiple sinks via a virtual root for the RPL routing protocol. EURASIP J. Wireless Commun. Netw. 2014(1), 91 (2014)
30. Vincze, Z., Vida, R., Vidacs, A.: Deploying multiple sinks in multi-hop wireless sensor networks. In: IEEE International Conference on Pervasive Services, pp. 55–63. IEEE (2007)
31. Safa, H., Moussa, M., Artail, H.: An energy efficient Genetic Algorithm based approach for sensor-to-sink binding in multi-sink wireless sensor networks. Wireless Netw. 20(2), 177–196 (2014)
32. Min, R., Bhardwaj, M., Cho, S.-H., Shih, E., Sinha, A., Wang, A., Chandrakasan, A.: Lowpower wireless sensor networks. In: International Conference on VLSI Design, pp. 205–210 (2001)
33. Min, R., Chandrakasan, A.: Energy-efficient communication for ad hoc wireless sensor networks. In: Signals, Systems and Computers, Conference Record of the Thirty-Fifth Asilomar Conference, vol. 1, pp. 139–143 (2001)
34. Anastasi, G., Conti, M., Falchi, A., Gregori, E., Passarella, A.: Performance measurements of motes sensor networks. In: Proceedings of the 7th ACM International Symposium on Modeling, Analysis and Simulation of Wireless and Mobile Systems (MSWiM'04), pp. 174–181 (2004)
35. Chang, J.H., Tassiulas, L.: Energy conserving routing in wireless ad-hoc networks. In: Proceedings of the 19th IEEE Conference on Computer Communications (INFOCOM), pp 22–31 (2000)
36. Kaurav, J., Ghosh, K.: Effect of transmitting radius, coverage area and node density on the lifetime of a wireless sensor network (2012)

The Design of Hierarchical Routing Protocol for Wireless Sensor Network

Ayan Kumar Das, Rituparna Chaki and Kashi Nath Dey

Abstract Energy efficiency is the main challenge for wireless sensor networks. Power-aware schemes send the same data through multiple paths, which causes data redundancy and a huge amount of energy drainage within the network. Use of a head node for a cluster can be a better solution. The proposed algorithm creates several clusters, selects a cluster head, aggregates the data, and sends that to base station through other cluster heads. Simulation result shows that it helps to increase network longevity.

Keywords Wireless sensor network · Cluster head · Network lifetime · Energy efficiency · Data aggregation

1 Introduction

The wireless sensor network is composed of a large number of sensor nodes that are densely deployed in a remote area and the base station or the access point. The sensor nodes produce some measurable responses to the changes in physical or chemical conditions. The main task of the sensor nodes is to sense those conditions

A.K. Das (✉)
Department of Computer Science and Engineering, Birla Institute of Technology,
Mesra, Off Campus Patna, India
e-mail: das.ayan777@gmail.com

R. Chaki
Department of A.K. Choudhury School of Information Technology,
University of Calcutta, Kolkata, India
e-mail: rituchaki@gmail.com

K.N. Dey
Department of Computer Science and Engineering, University of Calcutta,
Kolkata, India
e-mail: kndey55@gmail.com

© Springer India 2016   223
R. Chaki et al. (eds.), *Advanced Computing and Systems for Security*,
Advances in Intelligent Systems and Computing 395,
DOI 10.1007/978-81-322-2650-5_14

or a particular event and send the sensed data or the response to the base station or sink. It is quite difficult to recharge the battery of the nodes in this type of network as the nodes are deployed in a remote place. On the other hand, to fulfill the application requirements, the network lifetime should be long enough. Thus, energy-efficient routing is the main research area in wireless sensor network. The routing method followed by flat protocols like directed diffusion [1], rumor routing [2], minimum cost forwarding [3] etc. or the multipath-based routing protocols like GEAR [4], MERCC [5] wastes a huge amount of energy in data routing to the base station. Researches has shown that hierarchical protocols like LEACH [6], LEACH-C [7], PEGASIS [8] etc. are more energy efficient in this regard. Thus, the proposed protocol, can be abbreviated as HRP, takes a hierarchical approach in meeting the demand of energy-efficient routing and prolonging the network lifetime. In HRP, the cluster formation in done by the base station. This is followed by cluster head selection, sensing of data, and routing of data. The role of cluster head is rotated among the nodes of a cluster to distribute the load evenly among all the nodes in the cluster. Only the cluster heads of a particular round can take part in routing the data to the base station.

2 Related Work

Energy efficiency and increasing network longevity is the main research area in wireless sensor network for the last few years. Many algorithms are designed to form clusters and select the cluster head which has the responsibility to send the aggregated data to the base station to reduce the energy drainage. Some of them are discussed below.

Low energy adaptive clustering hierarchy (LEACH) [6] is a homogeneous clustering algorithm. It treats all the nodes without discrimination and forms clusters with only single-hop nodes based on the received signal strength and cluster heads are selected randomly. The role of cluster head is rotated among the nodes of the cluster to increase the network lifetime. Aggregated data is directly sent to the base station by each cluster head. In energy-efficient clustering algorithm for data aggregation in WSN (EECA) [9] two phases of clustering are mentioned. Formation of cluster is the first phase. It includes the broadcasting of messages to the neighbor nodes, advertising the radius of each node, residual energy and coordinates. The cluster head selection is based on competition bids calculated by each node. The selected cluster head broadcasts message of itself being the head. The next phase is data aggregation and tree construction. This phase includes calculation of weight on the basis of distance from the base station and then broadcasting of that weight thereby bringing energy efficiency, prolonged lifetime and reduced overhead. Clustering algorithm based on cell combination

(CACC) [10] proposed a clustering algorithm in which the monitoring region is divided into hexagonal cells. The division is done on the basis of geographical location of the nodes. Nodes in each cluster are given a particular cluster identity. Each cluster is divided into at least seven hexagonal cells with the cluster head selected from the central cell of each cluster such that the cluster head can easily communicate with nodes of any cell of the cluster. Clustering and multi-hop routing with power control in wireless sensor network [11] provides clustering together with power control. In set-up phase, the node with the maximum residual energy and maximum intracluster broadcast power, is chosen as the cluster head. In steady state phase, packets of request are sent by the base station confirming that those packets never appear twice for a node. On receiving the request, the sensor nodes send sensed data to the cluster head in a single hop. The nodes in turn follow a reverse way to send the aggregated data to the base station. The intracluster communication process of the algorithms is same as LEACH whereas intercluster communication takes place only between two cluster heads. The base station follows a power-efficient path to send requests to the cluster head. However, following the reverse path by the cluster heads leads to the consideration of the same nodes, which reduces their energy considerably. Maximum expected covering location problem (MXCLP) [12] points the necessity of uniform distribution of cluster heads throughout the network. This saves communication cost as well as prolongs the network lifetime. This algorithm uses a math programming approach which is based on the variation of a maximal expected covering location model. In this algorithm, the service failure of links are taken into account and the cluster heads are considered to be reliable. The mobile sensors form overlapping clusters with each node being assigned a particular cluster head. HEED [13] is the modified version of LEACH [6]. It is a distributed clustering scheme in which cluster head is selected on the basis of high residual energy. In this protocol, the probability of two nodes within each other's transmission range to become cluster head is small. The energy consumption for all the nodes in the network is not considered as same. In HEED, each node is mapped to exactly one cluster and can directly communicate with its cluster head. Proper energy dissipation is obtained in HEED. However, ignoring the network structure brings about uneven energy consumption throughout the network. Mobility-resistant efficient clustering approach (MRECA) [14] functions in a similar manner as that of HEED, with minimal complexities. Nodes are unaware of their locations in this algorithm. The local information and the respective score value in the network are broadcasted by the cluster heads. The score is used to calculate the delay for cluster head announcement. This algorithm is based on the mobility-resistant technique involving deterministic time without iterations. Better energy efficiency is ensured by this approach. The main feature is increased speed of clustering and robustness of the network. MRECA determines when the routing decision is made by the nodes. However, it does not focus on how it is performed.

The intercluster communication is not considered in this algorithm which adds to the drawback of this algorithm. In energy-efficient heterogeneous clustered scheme for wireless sensor network [15] it is assumed that a percentage of sensor nodes are equipped with more energy and are immobile with known geographical locations. The introduction of computational heterogeneity, which includes the presence of more powerful microprocessor, more energy and complex data processing ability together with longer term storage, added a lot of advantages to this model. The link heterogeneity is introduced with the inclusion of high bandwidth and long-distance network transceiver which prolonged the lifetime of the network together with reliable data transmission. The energy heterogeneity brought about the energy efficiency to the network, however increasing the implementation cost. In PEGASIS [8] nodes are arranged into chains and the communication takes place only with their closest neighbor. This kind of communication minimizes the power requirement for transmission of data. However, failure of any intermediate node can cut off the link between other nodes. Maximizing network lifetime through varying transmission radii with energy-efficient cluster routing algorithm in wireless sensor networks [16] proposes an improved LEACH [6] protocol for data gathering and aggregation. In the set-up phase of the algorithm, the sensor nodes identify the cluster to which they belong on the basis of the strength of the radio signal they receive from the cluster heads of that round. The cluster heads are selected on the basis of probability and required percentage of cluster heads of the network. A homogeneous network is assumed here and each noncluster head node send the data to the cluster head which in turn forward the data to the EECL node or the energy-efficient cluster head node of the network. The EECL node is the cluster head nearest to the base station. The node being closer to the base station dissipates less energy in transmitting the data to the base station thereby increasing the network lifetime. However, the residual energy of the EECL node is reduced considerably due to receiving of data from all the cluster head nodes and performing data aggregation. QOS supporting and optimal energy allocation for a cluster-based wireless sensor network [16] states that together with energy efficiency, the quality of service, which includes source to link delay, data pass rate, data loss rate etc. must also be taken under consideration. The algorithm states that each cluster is controlled by a cluster head having a finite capacity called single fixed rate. The relaying of traffic from cluster to cluster till the sink to minimize the data congestion and increase network lifetime, makes the cluster heads to depend on total relaying data rate from its own cluster as well as other clusters. Optimal energy allocation in heterogeneous wireless sensor network [17] mentions about sensors having different sensing range, different battery capacities. In this algorithm, nodes near the base station are equipped with more energy, and deployed in an arbitrary topology in an area such that they are connected. The paper provides calculation of probability distribution and expectation of the number of data transmitted in the lifetime of a sensor together with the probability distribution and expectation of the lifetime of the network. It also provides an algorithm design for optimal initial energy allocation to the sensors for prolonging the network lifetime coupled with the derivation of expected number of working sensors at each time. The algorithm takes into account the energy consumption during both sensing and transmitting data. The authors of [18] states that sensors within a cluster report their sensed data to their cluster head. That cluster head sends the aggregated data to higher level cluster head and so on until the data reaches the sink. The closer nodes from a cluster head is selected as next-level nodes by the sending cluster head. This selection procedure continues till the base station is reached. TEEN offers energy-efficient routing with better accuracy and dynamic response time. It combines hierarchical routing with data centric mechanism.

3 Proposed Work

3.1 Basic Method

In the proposed protocol HRP, it is assumed that the network is a static network, that is after deployment the nodes are motionless and contents of same initial energy. The base station or the sink node is assumed to be of high configuration. It is also assumed that all sensor nodes are capable of data aggregation, computing their own residual energy and finding their geographic location.

All the sensor nodes obtain their geographic location and send that to the base station. The base station groups the sensor nodes into different clusters based on their geographic location. Every cluster should have a cluster head to gather data from the other nodes of that region (cluster), aggregate that and send the aggregated data to the sink node. A huge amount of energy is needed for this type of transmission which may cause early death of the cluster head. Thus for every round, change of cluster head is required. The algorithm helps to select the new cluster head and route the aggregated data to the base station.

The energy required to send data from one node to another is proportional to distance and the total number of sent messages. Thus, energy consumed for a node i after sending some messages can be written as

$$E_{ci} = \sum_{j=1}^{n} k * \text{dist}_{ij} * N_m + E_a \qquad (1)$$

where, dist_{ij} is distance between nodes i and j, N_m is total number of sent messages from node i to j, k is a constant value, E_a is the energy required for data aggregation.

The residual energy for node i will be

$$E_{Res} = E_i - E_{ci} \tag{2}$$

where E_i is the initial energy.

If the residual energy E_{Res} of node i is greater than the threshold value, then it becomes a candidate node of cluster head selection process. Threshold value can be defined as energy required in accepting data from all nodes, aggregate that, and send that to neighbor nodes.

Every candidate node will calculate the selection probability Prob$_i$ value for itself. The probability value is given as

$$\text{prob}_i = E_{Res}/D_i \tag{3}$$

where

$$D_i = \sum_{j=1}^{n} \text{dist}_{ij}/n \tag{4}$$

where n is the number of neighbor nodes of node i.

Now all the nodes broadcast their probability value Prob$_i$ to the neighbors. Every node checks their own value, compares that with the others, and finds the maximum. The node containing the maximum Prob$_i$ value will send the success message to all its neighbors.

After selection of the cluster head, the aggregated data should reach to the base station in an energy-efficient way. The algorithm takes care of this by selecting a leader node among the cluster heads. To choose a leader, every cluster head will calculate a weight value by

$$W_i = k * \left(\frac{E_{Res}}{D_b} \right) \tag{5}$$

where, E_{Res} is the residual energy of cluster head i and D_b is the distance of that cluster head from the base station.

Each cluster head will broadcast its weight value to other cluster heads. The cluster heads compare the received weight values and the cluster head having the maximum weight value is considered as leader node. The leader aggregates all data coming from different cluster heads before sending it to the base station.

3.2 Algorithm

Input Node_locn [x,y] – the geographic location coordinates for each node.
Call clusterform (node_locn[x,y]) for forming the cluster depending on the value of x and y
Broadcast ClustInfoPct [Node ID, Cluster ID] to every node N_i.
    //Calculate energy consumed for every node i—
for i = 1 to n do-
    Repeat for j=1 to k
        a) Read Nm and Ea
        b) Eci=Eci+(c*d[i][j]*Nm)+Ea
//End of Inner loop
    //Calculate Residual Energy
        Set E[i]=E[i]-Eci
//End of outer loop
    For i=1 to n repeat step 8
        If E[i] > E_{th} Then
            a) Set S[i]=1
            b).For j=1 to n repeat
                c)   Set D=D+d[i][j]
                //End of loop
            d) Set Di=D/n
            e) Set Prob[i]=E[i]/Di
        Else
            Set S[i]=0, Prob[i]=0
//End of For loop
    //Broadcast probability value of every node to its neighbor within cluster only
    Call Prob_Broadcast(Prob[ ],n)
.   //Find the max value among the received probability values
    Call max()
.   For i=1 to n do-
        If Prob[i] > max then Send success message to all other nodes within the cluster

4   Simulation Result

A network with 100 nodes is created to analyze the performance of HRP. The list of parameters is given in Table 1.

The initial power of every node is considered 500 J. The size of each packet of data is taken as 1 kB. EECA [9] is chosen for comparison of performance. In EECA, one node can be a member of more than one cluster as shown in Fig. 1, which causes data redundancy and leads to more energy drainage. In HRP, base

Table 1 Parameter list

Parameters	Description
MAC protocol	IEEE 802.15.4
Network size	100 nodes
Initial energy	500 J per node
Number of rounds	At least 6
Power consumption	Equivalent to packet size and distance

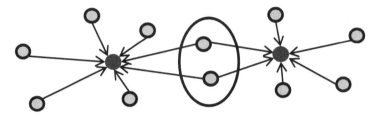

Fig. 1 Nodes in common field of two different cluster

station has formed the cluster so that no node can be there in common areas, thus reducing data redundancy and energy wastage.

In EECA, distance from the source node is not measured at the time of weight calculation. The energy depletion depends on distance and to get the optimized weight value both the distance, from the base station and from the source node should be measured. Consider Fig. 2 for example.

Here the values along with edges are representing the distance and by two paths aggregated data can be sent to the base station. Now we calculate the weight value according to Eq. (5) considering the fixed value $K = 5$ and $E_{Res} = 20$ for both the path.

Fig. 2 Sending aggregated data to base station through different path

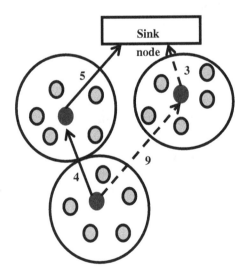

The weight value for the path with dotted line is 3.703 and that with the solid line is 5.00. Though the distance from the base station for the first one is smaller, it has the lower weight value due to the greater distance from the source node. Thus HRP makes uniform load distribution among the cluster heads and increases network lifetime by considering both the distance from the source node and from the base station.

In Fig. 3, it is observed that after completion of six rounds more nodes are dead for EECA than HRP.

Again after each round, the average residual energy of the network is calculated and plotted for EECA [9] and HRP as follows.

Figure 4, shows that after sixth round the algorithm EECA lost more energy than HRP. Thus, HRP is more energy aware than EECA.

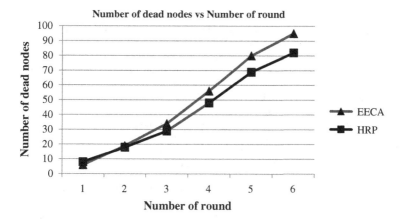

Fig. 3 Number of dead nodes versus number of round

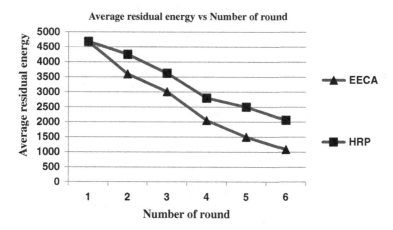

Fig. 4 Average residual energy versus number of round

5 Conclusion

The main concern of researchers in the domain of WSN is to provide energy-efficient routing due to the restrictions on battery power of the sensors. In this paper, an attempt has been made to provide efficient power utilization during routing of information in WSN. The proposed algorithm divides the network into several clusters, selects a cluster head, aggregates the data of that cluster, and sends that to the base station through other cluster heads. The proposed HRP logic has been compared against EECA, and the simulation result shows that network longevity is increased in case of HRP.

References

1. Intanagonwiwat, C., Govindan, R., Estrin, D., Heidemann, J.: Directed diffusion for wireless sensor networking. IEEE/ACM Trans. Netw. **11**, 2–16 (2003)
2. Braginsky, D., Estrin, D.: Rumor routing algorithm for sensor networks. In: the Proceedings of the First Workshop on Sensor Networks and Applications (WSNA), Atlanta, GA, Oct 2002
3. Ye, F., Chen, A., Liu, S., Zhang, L.: A scalable solution to minimum cost forwarding in large sensor networks. In: Proceedings of the Tenth International Conference on Computer Communications and Networks (ICCCN), pp. 304–309 (2001)
4. Yu, Y., Estrin, D., Govindan, R.: Geographical and energy-aware routing: a recursive data dissemination protocol for wireless sensor networks. UCLA Computer Science Department Technical Report, UCLA-CSD TR-01-0023, May 2001
5. Das, A.K., Dr. Chaki, R: MERCC: multiple events routing with congestion control for WSN. ACITY (1), vol. 176, pp. 691–698 (2012)
6. Pawa, T.D.S.: Analysis of Low Energy Adaptive Clustering Hierarchy (LEACH) protocol. National Institute of Technology, Rourkela (2011)
7. Geetha, V.A, Kallapurb, P.V, Tellajeerac, S.: Clustering in wireless sensor networks: performance comparison of LEACH & LEACH-C protocols using NS2. Elsevier J. (2012)
8. Lindsey, S., Raghavendra, C., Sivalingam, K.M.: Data gathering algorithm in sensor networks using energy metrics. IEEE Trans. Parallel Distrib. Syst. **13**(9) (2002)
9. Sha, C., Wang, R., Huang, H., Sun, L.: Energy efficient clustering algorithm for data aggregation in wireless sensor networks. Nanjing University of Posts and Telecommunications, Nanjing 210003, China. The Journal of China Universities of Posts and Telecommunications, January 2011
10. Chang-RI, L., Yun, Z., Xian-ha, Z., Zibo, Z.: A clustering algorithm based on cell combination for wireless sensor networks. In: Second International Workshop on Education Technology and Computer Science, vol. 2, pp. 74–77
11. Guo, S., Zheng, J., Qu, Y., Zhao, B., Pan, Q.: Clustering and multi-hop routing with power control in wireless sensor networks. J. China Univ. Posts Telecommun. **14**(1) (2007)
12. Feng, Y., Zhang, W.: A clustering algorithm for wireless sensor network. Int. J. Intell. Eng. Syst. **3**(4) (2010)
13. Kour, H., Sharma, A.K.: Hybrid energy efficient distributed protocol for heterogeneous wireless sensor network. Int. J. Comput. Appl. **4**(6), 0975–8887 (2010)
14. Li, Jason H., Miao, Yu., Levy, Renato, Teittinen, Anna: A mobility-resistant efficient clustering approach for ad hoc and sensor networks. ACM SIGMOBILE Mob. Comput. Commun. Rev. **10**(2), 1–12 (2006)

15. Kumar, D., Aseri, T.C., Patel, R.B.: Energy efficient heterogeneous clustered scheme for wireless sensor network. J. Comput. Commun. Elsevier **32**(4), 662–667 (2009). doi:10.1016/j.comcom.2008.11.025
16. Venu Madhav, T., Sarma, N.V.S.N.: Maximizing network lifetime through varying transmission radii with energy efficient cluster routing algorithm in wireless sensor networks. Int. J. Inf. Electron. Eng. **2**(2) (2012)
17. Li, K., Li, J.: Optimal energy allocation in heterogeneous wireless sensor networks for lifetime maximization. J. Parallel Distrib. Comput. Elsevier **72**(7), 902–916 (2012)
18. Manjeshwar, A., Agrawal, D. P.: TEEN: a protocol for enhanced efficiency in wireless sensor networks. In: Proceedings of the 1st International Workshop on Parallel and Distributed Computing Issues in Wireless Networks and Mobile Computing, San Francisco, CA, April 2001

Network Selection Using AHP for Fast Moving Vehicles in Heterogeneous Networks

Raman Kumar Goyal and Sakshi Kaushal

Abstract In today's world, there are various means of accessing the Internet such as cellular, wireless local area network (WLAN) and worldwide interoperability for microwave access (WiMAX), etc., for the mobile users. Also, various applications demand different quality of service (QoS) parameters. But for seamless connectivity in the case of fast moving vehicles, velocity of vehicle becomes an important issue. Traditional schemes trigger the handover process based on signal strength. These schemes do not incorporate the network parameters and user preferences required for optimal vertical handover. In this paper, analytic hierarchy process (AHP) method has been used for network selection in heterogeneous environments for moving vehicles. The method has been applied for various types of applications like conversational, streaming, interactive, and background applications. From the results, it has been found that WLAN's performance degrades significantly when the vehicles are moving at higher velocities while Universal Mobile Telecommunication Systems (UMTS) performs best for fast moving vehicles.

Keywords Mobility · Heterogeneous networks · AHP · Vertical handover · Network selection

1 Introduction

With the growth of wireless technology, Internet can be accessed by small hosts. Therefore, the demand of maintaining the Internet connectivity on move is arising. When a user moves from one place to another, there may be signal or other QoS

R.K. Goyal (✉) · S. Kaushal
Computer Science and Engineering, UIET, Panjab University, Chandigarh, India
e-mail: raman.uiet1985@gmail.com

S. Kaushal
e-mail: sakshi@pu.ac.in

© Springer India 2016
R. Chaki et al. (eds.), *Advanced Computing and Systems for Security*,
Advances in Intelligent Systems and Computing 395,
DOI 10.1007/978-81-322-2650-5_15

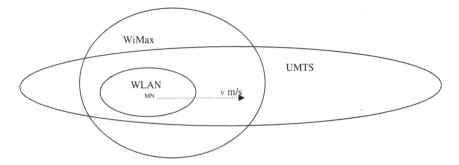

Fig. 1 Vertical handover scenario in a moving vehicle

degradation. So, user wants to connect to another network that provides better services. The process of switching from one network to another network is called handover. When this switching happens in same type of networks, it is called horizontal handover. Mobile devices have multiple interfaces to access the different technologies which tends the heterogeneous network system toward convergence. Vertical handover can be defined as the process of maintaining the connection while moving from one technology network to another [1]. Vertical handover occurs in heterogeneous networks. Selecting the network for vertical handover for different types of applications like streaming, interactive video conferencing, Voice over IP (VOIP), and background (browsing, etc.) require different set of network parameter values. When users are traveling in a moving vehicle, they may be in the range of different types of networks as presented in Fig. 1. While making vertical handover decision in a moving vehicle, velocity of the vehicle should be considered as an important decision criteria, as small range networks like WLAN do not support high velocities. In this paper, AHP-based network selection technique for heterogeneous networks is analyzed for various applications with having velocity of vehicle as the critical attribute, i.e., while assigning the weights using AHP, it has the maximum importance.

There may be different technologies available to the user for wireless connectivity and choosing the best alternative is the main objective of this paper. This selection procedure requires different parameters like data rate, delay, jitter, and the cost of network. In case of connectivity inside a vehicle, velocity also becomes a crucial parameter. As when the velocity increases, the performance of the handover procedure gets affected [2].

The rest of the paper is organized as follows. Section 2 presents the related works. The AHP technique for network selection is discussed in Sect. 3. Results of the approach are shown in Sect. 4. Finally, the conclusion is presented in Sect. 5.

2 Related Works

This section gives an overview of various multiple attribute decision-making (MADM) selection techniques used for network selection in heterogeneous networks. Many MADM techniques for decision making have been presented in [3]. Various researchers have conducted various studies for applying MADM techniques for network selection. Falwo et al. [4] presented technique for order preference by similarity to ideal solution (TOPSIS)-based network solution considering maximum data rate, security, delay, battery power consumption, and cost. This method selects the network based on single application or multiple applications. Song et al. [5] presented an AHP and Gray rational analysis (GRA)-based technique where AHP is used for selection criteria and GRA evaluates heterogeneous networks. Mohamed Lahby et al. [6] have proposed an enhanced TOPSIS method by calculating the relative closeness to the ideal solution by considering both the relative importance values of anti-ideal solution and ideal solution respectively. Liu et al. [7] projected fuzzy TOPSIS-based network selection for heterogeneous networks based on connection numbers. Lahby et al. [8] also presented the novel validation approach for network selection by considering the group weighting techniques for MADM methods. Zhang et al. [9] have applied MADM for network selection using AHP and synergetic theory. First, weights are calculated using AHP method and then the subsystem order degree is obtained. Four subsystems are considered, i.e., timeliness (delay and jitter), throughput (available data rate, maximum data rate), reliability (packet loss) and cost. Then entropy for the whole system is calculated. Based on this system entropy network is selected.

All the above-mentioned techniques considered various attributes like data rate, delay, jitter, cost, security, battery consumption, or energy, etc. But these methods ignore mobility of mobile device which is an important attribute while accessing Internet inside vehicles. Some researchers have tried to incorporate velocity of the device while taking network selection decision. Abolfazi et al. [10] have incorporated velocity in fuzzy extension to VIKOR for network selection. But velocity of the device has been varied for 0–10 m/s, which is not appropriate for fast moving vehicles. Dhar et al. [11] have also discussed mobility-based AHP method for selecting the network. Also, an intelligent system based on neural networks has been presented.

In this paper, we have demonstrated the affect of mobility in four traffic classes. Mobility is calculated as 10-point scalability based on the different ranges that a specific network can support. Detailed network selection scheme is presented in Sect. 3.

3 Proposed AHP Based Network Selection Method

AHP method was proposed by Saaty [12]. We have used AHP for network selection in fast moving vehicles. The network selection is based on five attributes namely data rate, mobility, cost, delay and jitter. Three networks are considered for network selection, i.e., WLAN, WiMAX and UMTS. Four types of applications are considered namely, conversational, interactive, streaming, and background. We assumed that the maximum mobility that a network can provide is 15, 60 and 100 m/s for WLAN, WiMAX and UMTS, respectively. Mobility is calculated on 10-point scale as follows 10–9*(velocity/maximum mobility of network). The AHP process for the network selection process is as follows.

Step 1: Determine the objective and evaluation parameters. Select the attributes and alternatives. In our problem following are the attribute values correspond to different types of networks as shown in Table 1.
Obtain the normalized matrix by dividing with the value of beneficial attribute (data rate, mobility) and dividing the non-beneficial attribute (cost, delay, and jitter) with the value of attribute.
Step 2: Construct a paired comparison matrix using a scale of relative importance. An attribute compared with itself is given a value of 1 and the values 3, 5, 7, and 9 corresponds to moderate importance, strong importance, very strong importance, and absolute importance. While, 2, 4, 6, and 8 compromise between these values. Relative importance matrices for different type of applications are shown in Tables 2, 3, 4 and 5 respectively. We name the matrix as $B1_{5*5}$ where b_{ij} denotes the relative importance of ith attribute with jth attribute. For conversational applications, bandwidth and cost requires less data rate as compared to delay and jitter. Interactive applications like video conferencing require higher data rate and lower delay. For video streaming and background applications, data rate has given more importance than delay a jitter. Delay attribute is more important to video streaming than in background applications.
Step 3: Find the relative normalized weight (w) for each attribute by calculating the geometric mean (GM) of the each row in the comparison matrix and normalize the geometric means of rows as follows

$$\text{GM}_j = \left[\prod_{j=1}^{M} b_{ij} \right]^{1/M}$$

Table 1 Network selection metrics for candidate networks

Network	Data rate (Mbps)	Mobility (m/s)	Cost	Delay (ms)	Jitter (ms)
WLAN	50	10–9*(v/15)	1	110	10
WiMAX	25	10–9*(v/60)	3	100	3
UMTS	2	10–9*(v/100)	5	35	5

Table 2 Relative importance of different attributes in conversational applications

	Data rate	Mobility	Cost	Delay	Jitter
Data rate	1	0.11	1	0.14	0.20
Mobility	9	1	9	3	5
Cost	1	0.11	1	0.33	0.33
Delay	7	0.33	5	1	2
Jitter	5	0.2	5	0.5	1

Table 3 Relative importance of different attributes in interactive applications

	Data rate	Mobility	Cost	Delay	Jitter
Data rate	1	0.33	5	1	2
Mobility	3	1	9	3	5
Cost	0.20	0.11	1	0.2	0.33
Delay	1	0.33	5	1	2
Jitter	0.50	0.2	5	0.5	1

Table 4 Relative importance of different attributes in streaming applications

	Data rate	Mobility	Cost	Delay	Jitter
Data rate	1	0.33	7	2	3
Mobility	3	1	9	5	7
Cost	0.14	0.11	1	0.33	0.5
Delay	0.5	0.20	5	1	2
Jitter	0.33	0.14	3	0.5	1

Table 5 Relative importance of different attributes in background applications

	Data rate	Mobility	Cost	Delay	Jitter
Data rate	1	0.33	7	3	3
Mobility	3	1	9	5	5
Cost	0.14	0.11	1	0.33	0.33
Delay	0.33	0.20	5	1	1
Jitter	0.33	0.2	5	1	1

and

$$w_j = GM_j / \sum_{j=1}^{M} GM_j$$

where M is the order of matrix. We name the w^T matrix as B3. GM and B3 matrices are shown in Tables 6 and 7 respectively.

Table 6 GM of each attribute for different applications

	Conversational	Interactive	Streaming	Background
Data rate	0.315838	1.269705	1.691814	1.834725
Mobility	4.139189	3.32269	3.936283	3.68011
Cost	0.413578	0.27115	0.303301	0.279115
Delay	1.873796	1.269705	1	0.80113
Jitter	1.201124	0.757858	0.586336	0.80113

Table 7 Relative normalized weight (w^T) for specific applications of each attribute

	Conversational	Interactive	Streaming	Background
Data rate	0.03976	0.184251	0.225043	0.248063
Mobility	0.521077	0.482169	0.5236	0.497567
Cost	0.052065	0.039353	0.040345	0.037738
Delay	0.23589	0.184251	0.133019	0.108316
Jitter	0.151208	0.109976	0.077994	0.108316

Table 8 B3 matrix for specific applications

	Conversational	Interactive	Streaming	Background
Data rate	0.213084	0.944333	1.180263	1.32632
Mobility	2.811213	2.49173	2.772882	2.664556
Cost	0.277465	0.20292	0.212922	0.19924
Delay	1.248908	0.944333	0.707971	0.586694
Jitter	0.832494	0.587424	0.413106	0.586694

Table 9 B4 matrix for specific applications

	Conversational	Interactive	Streaming	Background
Data rate	5.359187	5.125241	5.244609	5.34671
Mobility	5.395005	5.167753	5.295805	5.35517
Cost	5.329228	5.156433	5.277555	5.279632
Delay	5.294457	5.125241	5.322341	5.416494
Jitter	5.505625	5.341407	5.296651	5.416494

Table 10 λ_{max} for specific applications

	Conversational	Interactive	Streaming	Background
λ_{max}	5.3767	5.183215	5.287392	5.3629

Table 11 CR values for different applications

Application	Conversational	Interactive	Streaming	Background
CR	0.084842	0.041265	0.084842	0.081734

Step 4: Calculate B3 = B1*B2 and B4 = B3/B2 as shown in Tables 8 and 9 respectively.

Step 5: Calculate the maximum Eigen value λ_{max} of B4 matrix as shown in Table 10.

Step 6: Calculate the consistency index CI = $(\lambda_{max} - M)/(M - 1)$.

Step 7: Obtain the Random Index (RI) for the number of attributes used in decision making. For five attributes RI = 1.11.

Step 8: Calculate the consistency ratio CR = CI/RI. A CR of 0.1 or less is acceptable. From Table 11, it is evident that all the considered matrices are consistent.

Step 9: Calculate the overall AHP score by multiplying the normalized weight of the attribute calculated in Step 1 by w_j calculated in Step 3.

4 Results and Discussion

AHP score is calculated as discussed in Sect. 3. If this score becomes negative, the value is assumed to be zero. When the vehicle is stationary, WLAN provides maximum data rate and the attribute mobility has zero value. Therefore, it outperforms UMTS and WiMAX networks in case of interactive, streaming and background applications. While for conversational applications, UMTS network is selected as it provides lowest delay and jitter. Performance of these three networks for all the applications are shown in Fig. 2a.

As the vehicle's speed starts increasing, the value of mobility attribute becomes lesser. For WLAN, it has declined much earlier, as it can support maximum mobility up to 15 m/s. AHP score has become negative after 20 m/s, but it is represented as zero. The performance of WiMAX is also hampered in case of fast moving vehicles. AHP scores of WiMax are relatively good for velocities below 50 m/s. After 50 m/s, these scores start declining. Therefore, when the vehicles are moving fast, it is not considered as a good option. Apart from the conversational applications, UMTS is the last ranked choice for stationary users. This is due to the fact that UMTS provides least data rate among the three networks. But the mobility attribute's value in the normalized matrix obtained in Step 1 of the applied technique is always one. As it is the attribute having most relative importance, AHP scores for UMTS remain consistent. Performance of these networks for background, streaming, conversational and interactive applications are shown in Fig. 2b–e respectively.

Fig. 2 **a** AHP scores for the applications when the vehicle is not in motion. **b** AHP scores of networks for background applications. **c** AHP scores of networks for streaming applications. **d** AHP scores of networks for conversational applications. **e** AHP Scores of networks for interactive applications

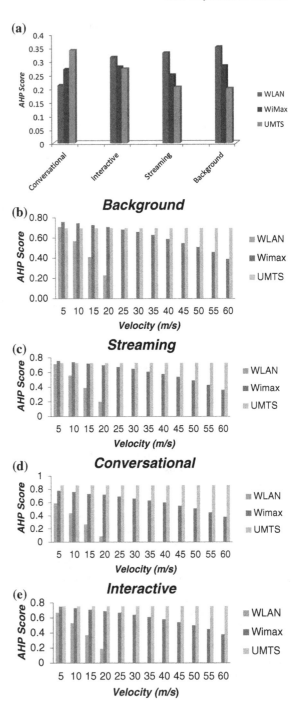

5 Conclusion

Seamless connectivity is the major concern while accessing Internet in moving vehicles. Short-range networks provide greater data rate but supports lesser mobility. In this paper, AHP has been used for network selection in case of fast moving vehicles for three different networks, i.e., WLAN, WiMAX, and UMTS. WLAN performs best for stationary vehicles, but as the vehicles start moving its performance degrades considerably. While performance of WiMAX is affected at high speeds, it performs well at lower and moderate speeds. UMTS is the most preferred choice when the Internet has to be accessed in fast moving vehicles. Further, work can be extended to anlayze the various other MADM methods on network selection and to optimize the weights given to the attributes for network selection criteria.

References

1. Budisz, L., Ferrus, R., Brunstorm, A., Grinnemo, K., Fracchia, R., Galante, G, Casadevall, F.: Towards transport-layer mobility: evolution of SCTP multihoming. Comput. Commun. **31**(5), 980–998 (2008)
2. Tao, M., Yu, H.: A smooth handover scheme for fast-moving users in mobile IPv6 networks. Wirel. Pers. Commun. **60**(4), 649–664
3. Rao, R.V.: Decision Making in Manufacturing Environment Using Graph Theory and Fuzzy Multiple Attribute Decision Making Methods. Springer, London (2007)
4. Falowo, O. E., Chan, H.A.: Dynamic RAT selection for multiple calls in heterogeneous wireless networks using group decision-making technique. Comput. Netw. **56**(4), 1390–401 (2012)
5. Song, Q., Jamalipour, A.: An adaptive quality-of-service network selection mechanism for heterogeneous mobile networks. Wirel. Commun. Mob. Comput. **5**(6), 697–708 (2005)
6. Lahby, M., Cherkaoui, L., and Adib, A.: An enhanced-TOPSIS based network selection technique for next generation wireless networks. In: 20th International Conference on Telecommunications, IEEE, pp. 1–5 (2013)
7. Liu, X., Chang, C.: The TOPSIS algorithm based on a + bi type connection numbers for decision-making in the convergence of heterogeneous networks. In: 3rd International Conference on Advanced Computer Theory and Engineering (ICACTE), pp. 323–327. IEEE Press, New York (2010)
8. Lahby, M., Cherkaoui, L., Adib, A.: Novel validation approach for network selection algorithm by applying the group MADM. In: ACS International Conference on Computer Systems and Applications (AICCSA), pp. 1–4. IEEE Press, New York (2013)
9. Zhang, L., Zhu, Z.: Multiple attribute network selection algorithm based on AHP and synergetic theory for heterogeneous wireless networks. J. Electron. (China), 29–40 (2014)
10. Mehbodniya, A., Kaleem, F., Yen, K.K., Adachi, F.: A fuzzy extension of VIKOR for target network selection in heterogeneous wireless environments. Phys. Commun. **7**, 145–155 (2013)
11. Dhar, S., Ray, A., Bera, R.: Cognitive vertical handover engine for vehicular communication. Peer Peer Netw. Appl. **6**(3), 305–324 (2013)
12. Saaty, T.L.: The Analytic Hierarchy Process. McGraw-Hill, New York (1980)

Part IV
Data Analytics

Context-Aware Graph-Based Visualized Clustering Approach (CAVCA)

K. Rajendra Prasad and B. Eswara Reddy

Abstract The Clustering algorithms cannot detect the number of clusters for unlabeled data. The visual access tendency (VAT) is recognized as the best approach for cluster detection. However, context-aware-based graphs (CAG) give more informative cluster assessment for VAT. Hence, we extend the VAT using CAG, known as CAVAT. This paper investigates the existing cluster detection methods and proposes a data clustering method for the CAVAT for archiving the efficient clustering results.

Keywords Clustering · VAT · Clustering tendency · Hypergraph

1 Introduction

Clustering techniques are widely used in many data mining applications such as data clustering, image segmentation, pattern recognition, video motion segmentation, etc. The fundamental nature of clustering is unsupervised because it classifies the unlabeled data without the knowledge of prior labels of data objects. Many of clustering algorithms such as k-means [1], hierarchical clustering concept [2], density-based clustering concept [3], and graph-based clustering [4] are investigated for cluster analysis. In such algorithms, the quality of clusters depends on the number of clusters (i.e., k value in k-means); hence, determining the number of clusters is required. The user poses a value to the variable of 'k'; however, it may be unsuitable in some practical cases. Therefore, finding the number of clusters (i.e., the clustering

K. Rajendra Prasad (✉)
Department of IT, Rajeev Gandhi Memorial College of Engineering and Technology, Nandyal, A.P., India
e-mail: krprgm@gmail.com

B. Eswara Reddy
Department of CSE, JNTUA College of Engineering, Ananthapuramu, A.P., India
e-mail: eswar.cse@jntua.ac.in

© Springer India 2016
R. Chaki et al. (eds.), *Advanced Computing and Systems for Security*,
Advances in Intelligent Systems and Computing 395,
DOI 10.1007/978-81-322-2650-5_16

tendency) is an emerging need for such clustering algorithms. After going the thorough survey of literature, we address the clustering tendency using visual access tendency (VAT) method [5] for improving the quality of clusters. Other VAT algorithms such as SpecVAT [6] and iVAT [7] are described for the better assessment of clustering tendency. The VAT is an automatic cluster detection tool, and it determines the number of clusters in a visual form; hence, it is known as a visual procedure.

The present clustering algorithms (such as k-means or any traditional clustering algorithms) require the clustering tendency (i.e., number of clusters as input) for the effective clustering. It uses two procedures for achieving the accurate clustering. These are as follows: (1) It is required to use VAT for determining the number of clusters, i.e., to determine 'k' value in k-means; and (2) It is needed to run the classical k-means for clustering results. However, it is expensive from execution point of view. Hence, this paper is to address this problem in the proposed method. This proposed method uses two vital steps for improving the clustering method, which are as follows: (1) improve the VAT method using context-aware dissimilarity features (it is called as a 'CAVAT'), since the context-aware dissimilarity features are useful for better assessment of clustering tendency, (2) to use visualized clustering approach (VCA) [8, 9] for retrieving the explicit clustering results directly from 'CAVAT.' This proposed clustering method is known as a context-aware-based visualized clustering approach (CAVCA).

The key steps of the proposed CAVCA are described as follows:

1. Construct the pairwise hypergraph and kNN hypergraphs for a set of objects.
2. Compute the context-aware-based dissimilarity matrix (CAD) using pairwise and kNN (k-nearest neighbors) hypergraphs [10].
3. Use CAD matrix in VAT for developing the CAVAT and to extract the clustering tendency effectively from CAVAT.
4. Develop the CAVCA using CAVAT for discovering the clustering results.

The procedure for context-aware-based dissimilarity matrix computation is described in [10]; it uses the k-nearest neighbor's concept for hyperedges detection, and it finds the dissimilarity matrix with respect to hyperedges [11]. The proposed CAVCA is an efficient method for clusters generation.

The rest of the paper is organized as follows: Sect. 2 presents the related work. Section 3 discusses visualized clustering approach (VCA). Section 4 presents the proposed method. Section 5 discusses the experimental study and results, and Sect. 6 presents the conclusion and future scope.

2 Related Work

Determining the number of clusters is one of the pre-clustering issues because the data is unlabeled. The clustering methods require the prior knowledge about the number of clusters 'c,' for achieving the quality of clusters, since the quality of clusters depends on the number of clusters. Determining the number of clusters of unlabeled data is known as the clustering tendency. The statistically based technique is developed in [12] for the assessment of clustering tendency. Mean shift (MS) [13] is a successful method in the area of speech clustering, and it also finds the modes (or the number of clusters) of unlabeled data in a nonparametric form. This procedure is also known as a mode-seeking procedure. The basic idea of MS is as follows. Let take 'S,' where 'S' denotes the unlabeled data, i.e., $S = \{x_1, x_2,.... x_n\}$, where '$n$' refers the total number of object and to select the subset $S_h(x)$ of n_x samples based on the following Eq. (1) [14]:

$$S_h(x) \equiv \{x_i : \|x_i - x\| \leq h\} \tag{1}$$

The clustering results of MS are implicitly depends on bandwidth 'h.' The selection of the h in MS is a critical task because 'h' is not a static variable and the value of 'h' depends on spread of the data. This is a key limitation of the MS procedure. A Bayesian information criterion (BIC) [15] is another modern approach for data segmentation. It is possible to estimate the number of clusters by BIC, but it is sensitive to a large dataset. The visual methods such as VAT, SpecVAT, and iVAT are investigated from [16] to the assessment of clustering tendency. The VAT is a clustering tendency tool and it is used for extracting the cluster count [17]. It is useful in exploratory data analysis. It aims to generate the potential clusters or groups in a visual manner. The VAT uses the pairwise dissimilarity information as the input and it reorders the indices of dissimilarity matrix by Prim's logic for obtaining of reordered dissimilarity matrix 'RDM.' The image of RDM is the output of VAT method and this image is known as a VAT Image. VAT Image [16] shows the hidden clusters in the form of square-shaped dark blocks along the diagonal. Figure 1 shows the example of VAT. The basic VAT algorithm is proposed by Bezdek [5].

The scalable VAT algorithm (sVAT) is implemented in [18] and it addresses the clustering assessment problems for large datasets. The coVAT [19] algorithm performs the assessment of clustering tendency for the co-clustering problem. For large datasets, bigVAT [20] is used for assessment of clustering tendency and it is another improved scalable version. It addresses the scalability problem of big datasets using distinguished features (DFs) of objects. The DFs are used for deriving the best sample data of 'Bigdata.' Most of the clustering applications involve complex datasets; so it is required to determine the clustering tendency for complex datasets. Wang et al. [21] have proposed the SpecVAT for determining the clustering tendency of complex

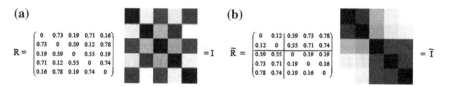

Fig. 1 VAT illustrative images. **a** Dissimilarity image 'I' (before applying the VAT) [16]. **b** After applying VAT [16]

datasets. The specVAT uses a spectral concept for reliable estimation of clusters. An improved VAT (iVAT) is recently proposed by Havens et al. [7]. The iVAT determines the clustering tendency for path-based datasets. The existing VAT algorithms cannot consider the influence of nearest neighbors during assessment of clustering tendency. Therefore, this paper addresses this issue in the proposed CAVAT. Another limitation of VAT methods is that they determine the clustering tendency; however, they cannot discover the clustering results. The CAVCA is proposed for addressing this issue. Hence, two key issues are addressed in this paper, which are as follows: to address that how to assess the clustering tendency of unlabeled data using context-aware dissimilarity-based VAT (CAVAT) and to address that how to extract the clustering results using VCA. The details of VCA are discussed in next section.

3   Visualized Clustering Approach (VCA)

The VCA [8] extends the VAT method for discovering the clustering results. The VCA consists of three vital steps and are described as follows: (1) Detect the clustering tendency from the VAT Image; (2) Obtain the crisp partition matrix from the VAT Image and this matrix defines the cluster labels of objects; and (3) Find the clustering results using crisp partition matrix. The current labels of objects are compared with ground truth labels for the purpose of finding clustering tendency. Algorithm 1 illustrates the VCA procedure [8].

Algorithm 1: VCA
Input   : D-Dissimilarity Matrix
               N- Number of objects
Output   : k- Number of Clusters

Method :
Step 1:
            [RD] =VAT(D)
            VAT_Im=Image(RD);
            k=No_of_Square_Shaped_Dark_Blocks(VAT_Im);
Step 2:
            For i = 1: k
            Find data objects at each partition i using crisp- partition matrix
            End for
Step 3:
            Map the data objects and find the ground truth labels using Khun-munkres function [27]
            for finding of clustering accuracy

The VCA is an enhanced visualized clustering approach; it is an extended version of VAT. The primary objective of the VAT is that it accesses the number of clusters from the VAT Image, for example, to obtain the number of clusters as two by counting the square-shaped dark blocks of VAT image in Table 1. However, the VAT detects only the number of clusters and may not discover the clustering

Table 1 Illustration of VAT and VCA procedures

Input variable	VAT	VCA
D—Sample dissimilarity matrix D = [0 0.73 0.19 0.71 0.16; 0.73 0 0.59 0.12 0.78; 0.19 0.59 0 0.55 0.19; 0.71 0.12 0.55 0 0.74; 0.16 0.78 0.19 0.74 0];	Reordered dissimilarity matrix RD = VAT(D) //*We obtain the RD from VAT algorithm*/ RD = [0 0.12 0.59 0.73 0.78; 0.12 0 0.55 0.71 0.74; 0.59 0.55 0 0.19 0.19; 0.73 0.71 0.19 0 0.16; 0.78 0.74 0.19 0.16 0]; The image of RD (known as VAT Image) is shown as	It is extended VAT i.e., it uses the steps of VAT and it extended by the following steps /* find k (number of clusters) and crisp partition matrix of VAT Image */ k = count_number_of_square_ shaped_dark_blocks(VAT Image); cp = crisp_partition_matrix (VAT_Image);For the given example, k = 3, and cp = [1 1 0 0 0; 0 0 1 1 1]; Therefore, the data objects of clusters are C1 = {o1*,o2*}; C2 = {o3*,o4*,o5*}; where * indicates the reordering
	VAT detects the clustering tendency (number of clusters)	VCA discovers the clustering results along with known clustering tendency

results. For this aspect, the VAT is extended as VCA in [8] for retrieving the best clustering results instead of using the traditional clustering approaches. The VCA determines the crisp partition matrix [8] from the RDI of VAT. The VCA determines the cluster labels of data objects from the crisp partition matrix. The following table illustrates the examples for VAT and extended VAT (known as visualized clustering approach—VCA) procedures.

The spectral concept is very familiar and robust approach to clustering because it used the graph analysis in various applications such as dimensionality reduction [22], video segmentation [23], and data clustering [23]. The spectral-based VAT algorithm was described in [21], known as SpecVAT. After the thorough analysis of SpecVAT, it is recommended to retrieve the data partitioning results using 'SpecVAT Image' instead of 'VAT image' in step 1 of Algorithm 2. Since the clarity of SpecVAT Image is better than VAT Image, the quality of clustering depends on the contrast of square-shaped block blocks. Figure 2 shows the clarity of block structures for both VAT and SpecVAT images for synthetic data. It is observed that the clarity of SpecVAT image is high. Hence, the SpecVAT is extended as SpecVCA for achieving the best clustering results.

The SpecVAT image shows more clarity of square-shaped dark blocks than VAT image. Based on the observation of spectral concept, it is strongly recommend to embed the dissimilarity features of objects into a k-dimensional spectral space, where 'k' refers to the number of Eigenvectors. The noteworthy steps of the SpecVAT [24] are as follows:

Fig. 2 VAT image and SpecVAT image for synthetic data [24]

VAT Image

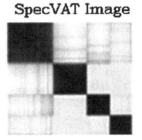SpecVAT Image

1. Compute the distance matrix 'W' for a set of objects using the local statistics of k-nearest neighbors, which results in maximum affinities within clusters and minimum affinities across clusters.
2. Construct the normalized Laplacian matrix 'L' as follows:

$$L = M^{(-1/2)}(M - W)M^{(-1/2)}$$

3. Choose the k largest eigenvectors of L and to form the eigen matrix $V = [v_1 v_2 \ldots \ldots v_K] \in R^{n \in k}$
4. Normalize the 'V' by Euclidean norm and this normalized V is denoted as 'V_1,' Find a new dissimilarity matrix 'Dnew' of V_1 with respect to n objects (of k-eigen dimensions).
5. Apply the VAT for the input 'Dnew,' to obtain the SpecVAT image.

The data visualized methods such as VAT, SpecVAT, and iVAT perform the assessment of clustering tendency without considering the nearest-neighbor influence. There is a scope to improve the existing VAT method with kNN hypergraph. In kNN hypergraph, the dissimilarity matrix is computed by the context of nearest neighbors; this idea is borrowed from [10].

4 Proposed Method

Similar objects share the common properties or common contexts. The existing clustering algorithms focused on the pairwise similarity between the vertices. If any one of the vertices is changed or removed, then the pairwise similarity for a set of vertices is significantly changed. Therefore, it may not be able to define a real affinity matrix. This is key issue of VAT (or SpecVAT). For handling this critical issue, context-aware dissimilarity-based VAT (CAVAT) is proposed for finding the stable clustering tendency and we extracting the clustering results from CAVAT image.

4.1   Context-Aware-Based Hypergraph Similarity Measures (CAHSM)

The CAHSM [10] explores the robust affinity relationships among vertices. The steps of CAHSM are given here: (1) construct a pairwise hypergraphs, and (2) find the k-nearest neighbor (kNN) hypergraph. Let the dataset $D = \{d_i\}_{i=1}^N$, with 'N' data objects. Create the 'N' number of vertices (is the set 'V') for a set of 'N' data objects of 'D.' Let the graph 'G' is denoted mathematically as $G = (V, E, W)$, $E \in V \times V$ denotes an edge set, and 'W' returns the affinity value between two vertices. The weighted similarity matrix of graph 'G' is defined by Eq. (2), and it is $A = (a_{ij})_{N \times N}$

$$a_{ij} = \begin{cases} W(v_a, v_b) \text{ if } (v_a, v_b) \in E \\ 0, \text{ otherwise} \end{cases} \tag{2}$$

In [11], the hypergraph analysis is introduced for capturing the underlying affinities among the vertices; it successfully solves the pre-clustering issues. The hypergraph is obtained from a generalized pairwise graph by composing hyper-edges. Each hyperedge defines a set of vertices that share some common properties. A set of hyperedges associated with the hypergraph incidence matrix 'H' and the values of $H = (h(v_a, e_k))_{|V| \times |E|}$ can be derived from Eq. (3)

$$h(v_a, e_k) = \begin{cases} 1, v_a \in e_k \\ 0, \text{ otherwise} \end{cases} \tag{3}$$

The pairwise similarity for a set of objects (for hypergraph) is computed from Eq. (4) [10]

$$u_{ij} = \sum_{e \in E} \text{weight}(e) h(v_i, e) h(v_j, e) = a_{ij} \tag{4}$$

Further, to construct the kNN hypergraph based on the nearest-neighboring information of vertices, i.e., to search the best k-neighbors of each vertex in kNN hypergraph construction, the nearest neighbors form the kNN hyperedge; it is denoted as $E = \{e_\ell^n\}_{\ell=1}^N$, where '$N$' is the number of nearest neighbors, and 'N' may probably be a small value (3 in our experiments). Use the equations (Eqs. 5–8 in this paper) of [10] for computing the kNN hypergraph similarity between two vertices.

$$I(v_i, e_\ell) = \begin{cases} 1, v_i \in e_\ell \\ 0, \text{ otherwise} \end{cases} \tag{5}$$

$$\delta_\ell = \frac{1}{|e_t|} \sum_{j \in \{r/v_r \in e_t\}} a_{tj} \tag{6}$$

$$h(v_i, e_\ell) = \frac{a_{\ell i \sqrt{I(v_i, e_\ell)v}}}{\sqrt{\sum_{t=1}^{N} \delta_t(v_i, e_\ell) a_{ni}^2}} \tag{7}$$

$$b_{ij} = \sum_{e \in E} \delta_\ell h(v_i, e) h(v_j, e) \tag{8}$$

4.2 Context-Aware Dissimilarity Matrix-Based Visualized Clustering Procedure (CAVCA)

The CAVAT is the best data visualization method because it uses the context-aware-based dissimilarity matrix (i.e., it considers the pairwise dissimilarity matrix of hypergraph, and kNN hypergraph) for finding the effective similarity features (or dissimilarity features), whereas the VAT uses the Euclidean-based dissimilarity matrix as the input. The CAVAT finds the dissimilarity matrix from two graphs, namely, pairwise hypergraph and kNN hypergraph. These two graphs calculate the dissimilarity features of 'n' objects using the local statistical information of k-nearest neighbors. The proposed CAVCA is a visualized clustering procedure and it extends the CAVAT. The algorithm of CAVCA is as follows.

Algorithm 2: CAVCA

Step 1 : Define hyperedges for a set of objects using k-nearest neighbor concept. Each hyperedge is a set of edges; those edges have connected toward to k-nearest neighbors from the source vertex

Step 2 : Hypergraph incidence matrix (H) is computed using the Eqn. (2)

Step 3 : S1=pairwise hypergraph similarity matrix using Eqn. (3)

Step 4 : D1=pairwise hypergraph dissimilarity matrix=1-S1;

Step 5 : S2= pairwise kNN hypergrah similarity matrix (from Eqn. (7))

Step 6 : D2=pairwise kNN hypergraph dissimilarity matrix=1-S2;

Step 7 : D= avg(D1,D2)

Step 8 : Clustering_Results_for_CAVCA=VCA(D)

The Algorithm 2 explains the steps for finding the context-based dissimilarity matrix using both pairwise hypergraph and kNN hypergraphs. It improves the clustering results by the contexts of nearest neighbors. The proposed CAVAT considers the contexts of k-nearest neighbors in dissimilarity matrix computation, which is an enhanced version of VAT. The proposed CAVCA discovers the clustering results effectively from CAVAT image. The CAVCA is outperformed than VCA because the CAVCA considers the local statistics of k-nearest neighbors.

5 Experimental Study and Results and Discussion

In this paper, various experiments are conducted for evaluating the CAVCA using various datasets such as synthetic dataset, real dataset, and audio dataset. The performance measures such as clustering accuracy (CA) and normalized mutual information (NMI) are used in the experimental study for demonstrating the effectiveness of proposed method.

5.1 Datasets Description

During the evaluation of our methods, this paper uses the comprehensive datasets. The datasets are as follows: four synthetic datasets (S-1 to S-4 in Fig. 3), and eight real datasets (R-1 to R-8) [25], including iris, voting, wine, and seeds, and speech datasets (R-5 to R-8) from [26]. The details of these datasets are presented in Table 2.

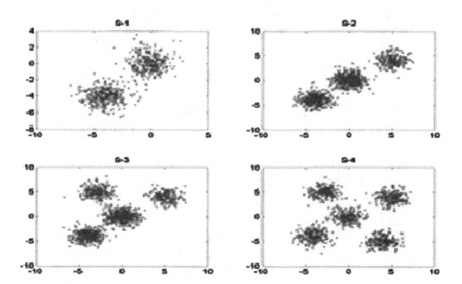

Fig. 3 Synthetic datasets (S-1 to S-4)

Table 2 Description of the datasets

Real dataset	Name of the dataset	No of clusters (or classes)
R-1	Iris	3 (No of classes)
R-2	Voting	2 (No of classes)
R-3	Wine	3
R-4	Seeds	3
R-5	Speech data from TSP	2 Speakers data
R-6	Speech data from TSP	3 Speakers data
R-7	Speech data from TSP	4 Speakers data
R-8	Speech data from TSP	5 Speakers data

5.2   Results and Discussion

The results of VAT images for the proposed methods are shown in Fig. 4a for S-4 synthetic data and Fig. 4b for speech data. From Fig. 4, it is investigated that only CAVAT image is appeared with more clarity than other VAT images; so it is more robust for determining the number of clusters. Each square-shaped dark block indicates a separate cluster. The size of the square-shaped dark block determines the

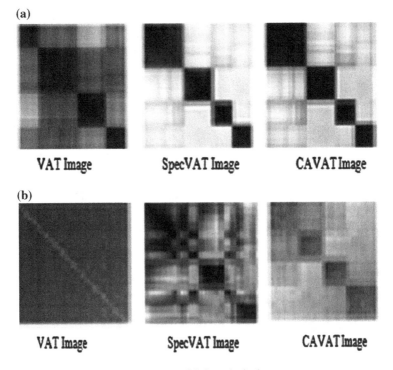

(a)

VAT Image   SpecVAT Image   CAVAT Image

(b)

VAT Image   SpecVAT Image   CAVAT Image

Fig. 4 Results comparison for R4 dataset and S-4 synthetic dataset

size of the cluster. The CAVAT uses the effective dissimilarity features, so that it enriches the clarity of square-shaped dark blocks. Excellent clarity and block structure of square-shaped dark blocks indicates the well-separated clusters. Hence, it is noted that the CAVAT is the best in practical assessment of clusters. In Fig. 4b, both VAT and SpecVAT (exiting algorithms) are unable to assess the clusters because the invisible square-shaped dark blocks are appeared. However, the CAVAT displays the visible and good contrasted square-shaped blocks along the diagonal. Therefore, it is easier to assess the clusters from CAVAT Image rather than VAT image (or SpecVAT).

5.3 Performance Evaluation and Comparison

The proposed methods are evaluated using three performance measures. These measures are clustering accuracy (CA) [27], goodness by OTSU, and normalized mutual information (NMI). The metrics of CA and NMI are widely used for evaluating the clustering performance. Assume that 'x_i^g' is ground truth label and 'f_i^g' is the clustering label, and then CA is defined as follows: $\max_{\text{map}} \sum_{i=1}^{n} \delta(x_i^g, \text{map}(f_i^g))/n1$, where $n1$ is the total number of objects. The function $\delta(x_i^g, \text{map}(f_i^g)) = 1$ if and only if $x_i^g = \text{map}(f_i^g)$, otherwise the value is 0. The sub-function 'map' permutes the clustering labels that match the equivalent labels given by the labels of ground truth. For finding this mapping, we use the Khun–Munkres algorithm to obtain a best mapping; the steps are described in [28].

The value of goodness is measured by OTSU concept [29], and Table 3 shows the 'goodness' of existing and proposed methods. The clustering accuracy and normalized mutual information are shown in Tables 4 and 5, respectively.

From the investigation of experimental results from Tables 3, 4 and 5, it is observed that CAVCA performs as the best, because the clustering accuracy and normalized mutual information values of CAVCA are high when compared to the existing VCA-VAT and VCA-SpecVAT methods.

Table 3 Performance measure: Goodness (by OTSU)

Real dataset	VCA-VAT	VCA-SpecVAT	CAVCA
S-1	0.9867	0.9961	**0.9956**
S-2	0.7622	0.8886	**0.8886**
S-3	0.7100	**0.8145**	0.8234
S-4	0.7100	**0.9980**	0.9976
R-1	0.8667	0.8511	**0.8733**
R-2	0.6345	0.6121	**0.6453**
R-3	**0.4264**	0.3251	0.3254
R-4	0.7135	0.7111	**0.7444**

Bold values indicate a maximum goodness

Table 4 Clustering accuracy (CA)

Real dataset	VCA-VAT	VCA-SpecVAT	CAVCA
S-1	0.4959	0.9384	**0.9444**
S-2	0.4981	0.9028	**0.9123**
S-3	0.4976	0.8625	**0.9222**
S-4	0.4986	0.8688	**0.8881**
R-1	0.3969	0.4060	**0.6401**
R-2	0.2931	0.7107	**0.7901**
R-3	**0.4899**	0.4058	0.4355
R-4	0.4581	0.4089	**0.6446**

Bold values indicate a maximum CA

Table 5 Normalized mutual information (NMI)

Real dataset	VCA-VAT	VCA-cVAT	VCA-SpecVAT	VCA-cSpecVAT
S-1	0.8978	0.0253	0.9666	**0.9678**
S-2	0.5249	0.3721	0.7641	**0.7644**
S-3	0.5934	0.5214	0.7081	**0.7092**
S-4	**0.9505**	0.5675	0.9922	0.9922
R-1	0.6563	**0.6963**	0.6265	0.6665
R-2	0.091	**0.1563**	0.0726	0.0856
R-3	**0.1529**	0.063	0.0113	0.0123
R-4	0.3197	0.1978	0.3214	**0.3215**

Bold values indicate a maximum NMI

6 Conclusion and Future Scope

The VAT is the data visualization method and it is used for the assessment of clustering tendency. The quality of clusters depends on the estimated value of clustering tendency. For this fact, this paper is mainly focused on VAT methods for finding the clustering tendency. The VCA is an existing clustering method; it discovers the clusters from VAT method. This paper improves the VAT as CAVAT using context-aware hypergraphs for the purpose of better assessment of clustering tendency. The VCA is also enhanced as CAVCA. The CAVCA extracts the clusters from the CAVAT. This CAVCA is an efficient clustering method and the effectiveness of proposed CAVCA clustering method is demonstrated in the experimental study. The scalability and processing time are challenging issues in Big Data clustering. Therefore, it needs to extend the proposed method in future for handing the Big Data issues using parallel processing approaches.

References

1. Kanungo, T., Mount, D.M., Netanyahu, N.S., Piatko, C., Silverman, R., Wu, A.Y.: An efficient k-means clustering algorithm: analysis and implementation. In: Proceedings of 16th ACM Symposium on Computational Geometry, pp. 1–21 (2000)
2. Jain, A.K., Murthy, M.N., Flynn, P.J.: Data clustering : a review. ACM Comp. Surveys **31**(3), 264–323 (1999)
3. Ester, M., Kriegel, P., Sander, J., Xu, X.: A density based algorithm for discovering clusters in large databases with noise. Int Conf. Knowl. Disc. Data Min. 226–231 (1996)
4. Zahn, C.T.: Graph theoretical methods for detecting and describing gestalt clusters. IEEE Trans. Comput. **20**(1), 68–86 (1971)
5. Bezdek, J., Hathaway, R.: VAT: A Tool for Visual Assessment (Cluster) Tendency, pp. 2225–2230. Proc. IJCNN, Honolulu (2002)
6. Wang, L. et al.: Enhanced visual analysis for cluster tendency assessment. Adv. Knowl Data Mining, LNCS, **6118**, 16–27(2010)
7. Havens, T.C., Bezdek, J.C.: An efficient formulation of the improved visual assessment of cluster tendency. IEEE Trans. Knowl. Data Eng. (2012)
8. Rajendra Prasad, K., Eswara Reddy, B.: An efficient visualized clustering approach for various datasets. IEEE SPICES 1–5 (2015)
9. Eswara Reddy, B., Rajendra Prasad, K.: Improving the performance of visualized clustering method. Int. J. Syst. Assur. Eng. (IJSA), Springer (2015)
10. Li, X., Hu, W., Shen, C., Dick, A., Zhang, Z.: Context-aware hyeprgraph construction for robust spectral clustering. IEEE Trans. Knowl. Data Eng. **26**(10) (2014)
11. Zhou, D., Huang, J., et al.: Learning with hypergraphs: clustering, classification, and embedding. P. NIPS (2006)
12. Dubes, R.C., Jain, A.K.: Clustering methodology in exploratory data analysis. In: Yovits, M. C. (ed.) Advances in Computers. Academic Press, New York (1980)
13. Ozertem, U., Erdogmus, D., et al.: Mean Shift spectral clustering. Patt. Recog. **41**(4), 1924–1938 (2008)
14. Senoussaoui, M., et al.: A study of the cosine distance-based mean shift for telephone speech diarization. IEEE Trans. Audio Speech Lang. Proc. **22**(1) (2014)
15. Tang, H., Chu, S.M.: Partially supervised speaker clustering. IEEE Trans. Pattern Anal. Mach. Intell. **34**(5), 959–971 (2012)
16. Wang, L., Nguyen, T., Bezdek, J., Leckie, C., Rammohanara, K.: iVAT and aVAT: enhanced visual analysis for clustering tendency assessment. Proc. PAKDD, India (Jun 2010)
17. Sledge, I.J., Huband, J.M., Bezdek, J.C.: (Automatic) Cluster count extraction from unlabeled data sets. In: Fifth International Conference on Fuzzy Systems and Knowledge Discovery IEEE Computer Society, pp. 3–13
18. Hathway, R., et al.: Scalable visual assessment of cluster tendency. Patt. Rec. **39**, 1315–1324 (2006)
19. Bezdek, J.C., et al.: Visual assessment of clustering tendency for rectangular dissimilarity matrices. IEEE Trans. Fuzzy Syst. **15**(5), 890–903 (2007)
20. Huband, J., et al.: Bigvat: visual assessment of clustering tendency for large datasets. Pattern Recogn. **38**(11), 1875–1886 (2005)
21. Wang, L., et al.: Enhanced visual analysis for cluster tendency assessment and data partitioning. IEEE Trans. Knowl. Data Eng. **22**(10), 1401–1413 (2010)
22. Belkin, M., Niyogi, P.: Laplacian Eigenmaps and spectral techniques for embedding and clustering. Adv. Neu. Inf. Proc. Sys. MIT Press (2002)
23. Wang, X., Mitchell, D.: A divide-and conquer approach for minimum spanning tree-based clustering **21**(7), 945–958 (2009)
24. Weiss, Y.: Segmentation using Eigen vectors: a unifying view. Proc. IEEE Int. Conf. Comput. Vision 975–982 (1999)
25. http://archive.ics.uci.edu/ml/datasets.html. (2002)

26. Kabal, P.: Audio File web document. http://WWW.TSP.ECE.McGill.CA/Docs/AudioFormats
27. Cai, D., He, X., Han, J.: Document clustering using locality preserving indexing. IEEE Trans. Knowl. Data Eng. **17**(2), 1624–1637 (2005)
28. Lovasz, L., Plummer, M.: Matching Theory. Budapest, Northholland (1986)
29. Otsu, N.: A threshold selection method from gray-level histograms. IEEE Trans. Syst. Man Cybern. **9**(1), 62–66 (1979)

Materialized View Construction Using Linearizable Nonlinear Regression

Soumya Sen, Partha Ghosh and Agostino Cortesi

Abstract Query processing at runtime is an important issue for data-centric applications. A faster query execution is highly required which means searching and returning the appropriate data of database. Different techniques have been proposed over the time and materialized view construction is one of them. The efficiency of a materialized view (MV) is measured based on hit ratio, which indicates the ratio of number of successful search to total numbers of accesses. Literature survey shows that few research works has been carried out to analyze the relationship between the attributes based on nonlinear equations for materialized view creation. However, as nonlinear regression is slower, in this research work they are mapped into linear equations to keep the benefit of both the approaches. This approach is applied to recently executed query set to analyze the attribute affinity and then the materialized view is formed based on the result of attribute affinity.

Keywords Materialized view · Non-linear regression · Curve fitting · Quantitative analysis · Hit–Miss ratio

S. Sen (✉)
University of Calcutta, Kolkata, India
e-mail: iamsoumyasen@gmail.com

P. Ghosh
Kingston School of Management & Science, KEI, Kolkata, India
e-mail: pghosh44@gmail.com

A. Cortesi
Universita Ca Foscari, Venice, Italy
e-mail: cortesi@unive.it

261

1 Introduction

Materialized view creation technique is used to create permanent views those are often used to answer the queries instead of using the base tables of the system. Materialized views are small in size compared to actual table and consist of important data; hence, faster query execution is possible. When the queries are generated in the system, materialized views are accessed first. If the results are not found in materialized views, then the actual tables of the system are searched. If data is available in materialized views, it is called hit and the failure is known as miss. The performance of a materialized view is measured in terms of hit ratio, which is computed as: hit/total access. Total access is computed as the sum of hit and miss. Materialized view is differentiated from ordinary view in terms of volatility. Views are not permanent in nature. Once the machines are switched off, the data is erased from the views, and again the data is loaded into it when it is called. Whereas materialized views are permanent in nature like tables. Hence no time is wasted in reloading the data every time. Thus materialized view is used mostly in large data-centric applications compared to other views.

Statistical analysis is a major approach to build materialized views. The result of the statistical analysis is presented in terms of a quantitative metric, which is used further to generate the materialized views. In this paper, the statistical analysis is performed using nonlinear regression. The theoretical benefits of nonlinear regression over linear regression are explained later in this paper. Moreover, the experimental results show better hit–miss ratio using nonlinear regression.

2 Related Work

The research on materialized views [1, 2] has been practiced over the years and all major database systems now support materialized views. Query Optimization [3] is one of the main focuses of creating materialized view. Materialized view is suitable for any data-related applications both large and small. Irrespective of the applications, materialized views focus on answering the users query directly from it. Even for the same application, the types of queries may be different for different types of users, hence based on the users requirements different materialized view can be created. A user-oriented approach SOMES [4] exactly meets this requirement. In order to support the dynamic nature of users' queries, dynamic materialized view creation methods are adopted in [5, 6]. These methods minimize the storage requirements and maintenance costs. Moreover it reduces searching time within the views, hence reduces the query answering time. Clustering-based dynamic materialized views [7, 8] techniques are another way of generating MVs. These methods at first define the clusters on MV and thereafter adjust the materialized view set dynamically. Automated selection [9] of materialized views is preferable to ensure dynamic changes. Popular algorithmic methods such as heuristic approach and genetic algorithms are also used in materialized views generation process. Greedy algorithm [2] based approaches were applied on "data cubes" to generate materialized views to optimize the query evaluation cost. Multiple view processing plan (MVPP) was proposed using heuristic algorithm [1, 10] to define an optimal materialized view selection scheme. The major focus of [1, 10] were (i) to achieve good performance and (ii) to ensure low maintenance cost. However, constraints related to storage were not considered. While creating the materialized views consideration of maintenance cost is important, AND-OR graph [11] based method helps to satisfy this consideration. Different constraints associated with materialized view creation are resolved using genetic algorithms [12, 13]. Storage space constraint is resolved in [13] using a genetic algorithm based solution, whereas [12] considers the materialized view creation for OLAP [14] applications under the storage space constraint and maintenance overhead. Outer join [15] helps higher query execution speed, and foreign key or referential integrity constraints reduce the maintenance cost. This survey work covers various techniques of materialized view creations. Some of the approaches directly work with the SQL queries to improve the searching capability in materialized view. Data warehouse based applications considers group by (roll-up) as an important operation. An approach for OLAP applications known as group query based on materialized view (GQMV) [6] enhance the searching speed by creating the materialized view for data warehouse applications which follows star schema.

Statistical analysis based on the attribute dependencies are analyzed for distributing the data in different locations in the case of distributed database. In order to incorporate the statistical analysis in materialized view creation process, a numeric scale [16] was proposed to analyze the dependencies between the attributes in a query set. The relationship among the attributes was identified using standard deviation [16]. However, in [16], materialized view creation process was not defined. In another approach these association among attributes are measured using linear regression [17]. The work also describes a methodology to create the materialized views. Although according to mathematical interpretation linear regression is stronger than standard deviation, both [16, 17] are linear by nature. However, the scientific or physical processes/data are generally nonlinear. The use of nonlinear approaches hence correlates the data or process in a more robust way than linear approaches. This motivates the researchers to apply nonlinear approach to create MVs.

The contribution can be seen as:

(a) To incorporate the concept of nonlinear equations to identify the association between the attributes.
(b) Construction of a quantitative metric based on nonlinear mathematical model.
(c) Based on this new quantitative metric, the materialized view is finally constructed according to available space for the application. The concept of nonlinearity is achieved by curve fitting algorithm.

3   Why Is Nonlinear Regression Required in This Context?

Linear regression method [17] calculates attribute affinity on the every participating attribute through the equation $Y = a + bX$.

Y denotes the total appearance of attribute i.
X identifies the mutual dependency of other attributes on attribute i.

As mentioned earlier scientific or physical processes/data are in general inherently nonlinear. Hence Linear Regression performs poorly to identify relationships among attributes. Moreover in [17] modular difference is calculated to find the difference between two attributes. Hence if two attributes differ by $+d$ and $-d$, respectively, then the result would be the same and both the attributes are equally fitted with that attributes, while the attribute with $+d$ is more important.

In fact, the functions used in real-world modeling are often inherently nonlinear in nature. This shortcoming motivates the researchers to use nonlinear method to overcome this problem. Another major advantage of using nonlinear regression is that different mathematical or statistical functions could be used here. Moreover, it has the additional benefit of computing (i) prediction (ii) confidence, and (iii) calibration intervals. Although the probabilistic interpretation of the intervals, generally, is only approximately correct, still these intervals are suitable for real-life cases.

However, the complexity of nonlinear regression is higher than linear regression. Thus a useful interpretation of this could be identifying a nonlinear equation that could be linearized. Hence a nonlinear regression which has the features of linear regression is important in this context. Thus identification of a nonlinear regression that can be linearized and computing unknown parameters with good precision from the input data set is suitable here. Linearizable curves require fewer numbers of parameters for calculation of values than the polynomials. Hence these types of linearizable curves are theoretically more suitable than the polynomials. Thus proper choice of the curve to fit the data can lead to identifying better relationship among the data member of the data set.

The above requirement is represented here in terms of the following nonlinear regression (curve fitting) equation

$$y = ax^b \tag{1}$$

y denotes the specific attribute (for every attribute this equation separately executed)
x identifies the corresponding correlation with respect to attribute y.
Equation (1) can be linearized by taking log on both sides

$$\log_{10} y = \log_{10} a + b \log_{10} x \tag{2}$$

Let us assume $y = \log_{10} y$; $A = \log_{10} a$; $X = \log_{10} x$

The required equation now takes the form

$$Y = A + bX \tag{3}$$

The proposed methodology uses nonlinear regression, which is more suitable and closely related with the real-world data. Moreover, as described in this section, the proposed methodology does not consider the modular difference between two attributes as done in the work like [17].

4 Proposal of Linearizable Nonlinear Regression

The contribution of this research work is classified into two parts. Initially, attribute relationship scale is formed which comprises of five steps. Attribute relationship scale is a metric to measure the association among the attributes quantitatively. In the next part materialized view is created from that scale. This view construction process is a two-stage process.

(a) Construction of Attribute Relationship Scale This process is started on a set of queries. We assume 'N' and 'M' as numbers of queries and numbers of attributes respectively. Using this input query set a ($N \times M$) *attribute query required matrix* (AQRM) is formed. Whenever an attribute appears in the query, the corresponding cell of AQRM is 1, otherwise 0. The sum is taken from the 1 to Nth rows of each attribute and is stored in the newly inserted ($N + 1$)th row of AQRM. This ($N + 1$)th row of AQRM is also termed as one-dimensional array 'Total_use'. Hence modified AQRM is now (($N + 1$) × M) matrix.

In the next step based on the mutual dependencies of the attribute a ($M \times M$) matrix is computed and termed as *attribute interrelation matrix* (AIM).

Next a nonlinear regression (curve fitting) method (Eq. (1)) is separately applied to each of the M attributes. Further this nonlinear regression is linearized using Eq. (2). The required equation takes the form of (3) as shown in previous section. The normal equations are expressed as

$$\sum Y = nA + b \sum X \tag{4}$$

$$\sum XY = A \sum X + b \sum X^2 \tag{5}$$

In the above equations, 'n' represents total number of attributes and the constants 'A' and 'b' are dependent on the regression curve. Solving the values of 'A' and 'b', 'a' is computed as

$$a = \text{antilog}_{10}(A)$$

In the next step, best fit curve matrix (BFCM) is constructed. This is computed by applying the equation to each attribute to store the result of dependencies of other attributes. Thereafter, using the BFCM matrix, for each attribute, the deviation of other attributes is computed and finally the result is stored in attribute relationship matrix (ARM). A new column is inserted in attribute relationship matrix (ARM) to store the total deviation of each attribute. This overall method is described using the algorithm Attribute_Relationship_Scale.

Algorithm Attribute_Relationship_Scale
Begin
Step 1: /* This step is same as QARM_Computation of [14]. The result is stored in (N × M) *Attribute Query Required Matrix* (AQRM) */
Function AQRM_Calculation (M,N)
Step 2: /* This step computes mutual occurrences of attributes from the matrix AQRM. The result is stored in (M×M) *Attribute Interrelation Matrix* (AIM)*/
Function AIM_Calculation (AQRM)
Step 3: /* Curve-Fitting Equations (CFE) are computed based on the probability of occurrence of each attribute*/
Function CFE_Calculation (AIM)
Step 4: /* From curve fitting equations Best Fit Curve Matrix (BFCM) is constructed to show how other attributes fit on the curve. This is a (M×M) matrix */
Function BFCM_Calculation (Curve Fitting Equations)
Step 5: /* Using Best Fit Curve Matrix(BFCM), the deviation between the attributes is stored in a (M x (M+1)) matrix called Attribute Relationship Matrix(ARM). (M+1)th column stores the summation of deviation of each attribute.*/
Function ARM_Computation (BFCM)
End

Function AQRM_Calculation (M, N)

```
Start
Repeat I = 1..N
 Repeat J = 1..M
     If J^th attribute present in I^th query
     AQRM[I, J]=1
     Increment Total_use[J] by 1
     else
     AQRM[I,J] = 0
     endif
  end loop
 end loop
Repeat I = 1..M
   AQRM[ N+1,I ] = Total_use[I]
   end loop
```

Function AIM_Calculation (AQRM)

Start
  Repeat I=1..M
    Repeat J = 1..M
    AIM [I,J] = Total numbers of times I and J occur together in the set of N queries;
      end loop
    end loop
End AIM_Calculation

Function CFE_Calculation (AIM)

Start
$y = a\,x^b$
$\log_{10} y = \log_{10} a + b\log_{10} x$
Let $Y=\log_{10} y$, $A=\log_{10} a$, $X=\log_{10} x$
Now $Y= A + bX$
Repeat I=1..M
   XSum=0, YSum=0, XYSum = 0
   Repeat J=1..M
     if(I≠J)
                XSum = XSum + X_J
                YSum = YSum + Y_I
         XYSum = XYSum + $X_J Y_I$
       endif
     end loop

   Solving the following equations as given in (4) and (5)
   YSum = m.A + b.XSum
   $XYSum = A.XSum + b.XSum^2$
   obtain the values of A and b.
    $a = (10)^A$
   Using A and b best fit curve for I^{th} attribute is computed
   end loop
End CFE_Calculation

Function BFCM_Calculation (Curve Fitting Equations)

/* This function uses the all curve fitting equations of the previous step to find out how all other attributes are to be fitted. */

```
Start
Repeat I = 1..M
  Repeat J = 1..M
    BFCM[I,J]= a₁ * ( AIM[ I,J] ) ᵇ¹
  end loop
end loop
End BFCM_Calculation
```

Function ARM_Computation (BFCM)

```
Start
Repeat I =1..M
  total=0
  Repeat J = 1..M
    ARM[I,J]= BFCM[I,I] - BFCM[I, J]
    total=total + ARM[I , J]
  end loop
  ARM[I, J + 1]=total_deviation
End loop
End ARM_Computation
```

(b) **Generation of materialized views** In this stage materialized view is constructed using the one-dimensional array 'Total_use' and the matrix attribute relationship matrix (ARM). 'Total_use' helps to find the important attributes whereas ARM helps to identify the mutual dependency of attributes. A $(2 \times M)$ matrix named ranked attribute and relationship matrix (RARM) is generated. Initially the attributes are ordered/sorted according to the descending order of use from 'Total_use'. If tie occurs between some attributes, then it has to be resolved as follows: The attributes are sorted in the ascending order from ARM according to the attribute affinity. In RARM the attributes are organized from left to right as the importance reduces. Hence the attribute at extreme left is the most important one and at the extreme right is the least important. These above-mentioned steps are depicted in an algorithm named MV_Creation, which is described below.

Algorithm MV_Creation

Begin

Step 1:

/* This step forms a (2×M) matrix called Ranked Attribute and Relationship Matrix (RARM) from Total_Use array and the matrix ARM.*/

Function RARM_Computation(ARM, Total_Use)

Step 2:

/* Finally, the materialized views are created, by choosing the attributes one after one from RARM in descending order of importance. The inputs are numbers of views C and the size of each view. The views are denoted as $V=\{V_1, V_2,...,V_C\}$ and their size as $S=\{S_1, S_2,...,S_C\}$. Initially all items of V are NULL */

Function View_Construction (V, S, C)

End

Function RARM_Computation(ARM, Total_Use)

```
Start
    In the 1ˢᵗ row of RARM the value of 'Total_use' is
    stored;
    In the 2ⁿᵈ row of RARM the corresponding value of
    (M+1)ᵗʰ attribute of AAM is stored;
    Sort the RARM according to the descending order of
    Total_Occurence (1ˢᵗ row)'
    If tie occurs among some attributes in 1ˢᵗ row then
        perform an ascending order sorting for those
        attributes, based on Total_deviation (2ⁿᵈ Row);
End RARM_Computation
```

Function View_Construction (V, S, C)

/* Sizeof function finds the size of the view. Attr denotes the attribute selected for clubbing with the existing view*/

```
Start
Repeat I=1..C
    Select RARM(I) /*Attribute at Iᵗʰ position of RARM*/
    Vᵢ= RARM(I)
    While (Sizeof(Vᵢ) < Sᵢ)
        Attr=Attribute with Minimum(ARM(I)) such that
        those attributes are not already included in Vᵢ

        If there is tie between Attr then select the Attr with
highest Total_Use from AQRM
        If there is still a tie between Attr then select the
Attr with smallest Total_Deviation from ARM
        Vᵢ= Vᵢ U Attr
    End While
End Loop
End View_Construction
```

5 Illustration by Example

The proposed method is illustrated here in terms of a simple example. We start with the same example of [17], a query set consisting of 10 queries. The numbers of attributes in this query set is also 10, they are namely $A_1, A_2, \ldots, A_{10}$.

Execution of Algorithm Attribute_Relationship_Scale

Step 1: *Attribute query required matrix* (AQRM) is given in Table 1 along with the $(N + 1)$th row Total_Use.

Step 2: *Attribute Interrelation Matrix (AIM)* is constructed using the method AIM_Computation. It is shown in Table 2.

Table 1 AQRM

	A_1	A_2	A_3	A_4	A_5	A_6	A_7	A_8	A_9	A_{10}
$q1$	1	1	1	1	1	0	0	0	1	0
$q2$	1	0	0	1	0	1	1	0	0	0
$q3$	0	0	1	0	1	1	1	1	0	0
$q4$	1	0	0	0	1	1	0	0	1	1
$q5$	0	1	0	0	0	0	1	1	1	1
$q6$	0	0	1	1	0	1	0	0	0	1
$q7$	1	1	1	0	0	1	0	1	1	0
$q8$	1	1	1	0	1	0	0	0	0	1
$q9$	0	1	1	0	1	0	0	1	1	1
$q10$	1	0	0	1	0	1	1	0	1	1
Total_Use	6	5	6	4	5	6	4	4	6	6

Table 2 AIM

	A_1	A_2	A_3	A_4	A_5	A_6	A_7	A_8	A_9	A_{10}
A_1	6	3	3	3	2	3	2	1	4	3
A_2	3	5	4	1	3	2	1	3	4	3
A_3	3	4	6	2	4	3	2	3	3	2
A_4	3	1	2	4	1	3	2	0	2	2
A_5	3	2	4	1	5	2	1	2	3	3
A_6	3	2	3	3	2	6	3	2	3	3
A_7	2	1	2	2	1	3	4	2	2	2
A_8	1	3	3	0	2	2	2	4	3	2
A_9	4	4	3	2	3	3	2	3	6	4
A_{10}	3	3	2	2	3	3	2	2	4	6

Step 3: CFE_Calculation is performed in this step. Here the calculation of A_1 is shown

$Y = 6$ for A_1 (obtained from Total_Use). CFE is computed for A_1 and is shown in Table 3.

Solving 'A' and 'b' we get $b = 1.16129$, $A = 2.6129$

Therefore $a = \text{antilog}_{10}(2.6129) = 410.1097$

$A_i = (410.1097) \cdot (\text{AIM}(A_1, A_i))^{1.16129}$

Similarly solving others,

$$A_i = (554.498) \cdot (\text{AIM}(A_2, A_i))^{0.7317}$$

$$A_i = (134.555) \cdot (\text{AIM}(A_3, A_i))^{1.2581}$$

$$A_i = (412.472) \cdot (\text{AIM}(A_4, A_i))^{0.6154}$$

$$A_i = (617.5897) \cdot (\text{AIM}(A_5, A_i))^{0.814}$$

$$A_i = (19.3062) \cdot (\text{AIM}(A_6, A_i))^{1.7143}$$

$$A_i = (50.687) \cdot (AIM(A_7, A_i))^{1.1148}$$

$$A_i = (303.8785) \cdot (\text{AIM}(A_8, A_i))^{0.6207}$$

$$A_i = (87.337) \cdot (\text{AIM}(A_9, A_i))^{1.2353}$$

$$A_i = (119.371) \cdot (\text{AIM}(A_{10}, A_i))^{1.3846}.$$

Step 4: Using the above equation for each attribute how other attributes are fitted on it is computed, and this result is stored in best fit curve matrix (BFCM) as shown in Table 4.

Step 5: Calculating the deviation of each attribute with respect to others, the final result is stored in a matrix called attribute relationship matrix (ARM) as shown in Table 5.

Table 3 CFE calculation for A_1

X	Y	XY	X^2
3	6	18	9
3	6	18	9
3	6	18	9
2	6	12	4
3	6	18	9
2	6	12	4
1	6	6	1
4	6	24	16
3	6	18	9
$\sum X = 24$	$\sum Y = 54$	$\sum XY = 144$	$\sum X^2 = 70$

Table 4 BFCM

	A_1	A_2	A_3	A_4	A_5	A_6	A_7	A_8	A_9	A_{10}
A_1	3285.2	1468.8	1468.8	1468.8	917.2	1468.8	917.2	410.1	2051.5	1468.8
A_2	1238.8	1800.26	1529.06	554.5	1238.8	920.8	554.5	1238.8	1529.06	1238.8
A_3	536	769.8	1282	321.8	769.8	536	321.8	536	536	321.8
A_4	811	412.47	632	968.06	412.47	811	632	0	632	632
A_5	1510.35	1085.77	1908.87	617.59	2289.08	1085.77	617.59	1085.77	1510.35	1510.35
A_6	126.9	63.35	126.9	126.9	63.35	416.56	126.9	63.35	126.9	126.9
A_7	109.77	50.69	109.77	109.77	50.69	172.5	237.72	109.77	109.77	109.77
A_8	303.88	600.96	600.96	0	467.25	467.25	467.25	718.45	600.96	467.25
A_9	484.1	484.1	339.3	205.6	339.3	339.3	205.6	339.3	798.8	484.1
A_{10}	546.4	546.4	311.7	311.7	546.4	546.4	311.7	311.7	813.8	1426.7

Table 5 ARM

	A_1	A_2	A_3	A_4	A_5	A_6	A_7	A_8	A_9	A_{10}	Total deviation
A_1	0	1816.4	1816.4	1816.4	2368	1816.4	2368	2875	12337.7	1816.4	17936.8
A_2	561.46	0	271.2	1245.8	561.46	879.5	1245.8	561.46	271.2	561.46	6159.34
A_3	746	512.2	0	960.2	512.2	746	960.2	746	746	960.2	6889
A_4	157.06	555.59	336.06	0	555.59	157.06	336.06	968.06	336.06	336.06	3737.6
A_5	778.73	1203.31	308.21	1671.49	0	1203.31	1671.49	1203.31	778.73	778.73	9669.31
A_6	289.7	353.2	289.7	289.7	353.2	0	289.7	353.2	289.7	289.7	2797.83
A_7	127.95	187.03	127.95	127.95	187.03	65.22	0	127.95	127.95	127.95	1206.98
A_8	414.57	117.49	117.49	718.45	251.2	251.2	251.2	0	117.49	251.2	2489.84
A_9	314.7	314.7	459.5	593.2	459.5	459.5	593.2	459.5	0	314.7	3968.5
A_{10}	880.3	880.3	1115	1115	880.3	880.3	1115	1115	612.8	0	8594

Table 6 RARM

	A_6	A_9	A_3	A_{10}	A_1	A_2	A_5	A_7	A_8	A_4
Totalocc.	6	6	6	6	6	5	5	4	4	4
Total dev.	2797.8	3968.5	6889	8594	17926.8	6159.3	9669.3	1206.9	2489.8	3737.6

ARM is the final output of the algorithm Attribute_Relationship_Scale. Now algorithm MV_Creation would be executed.

Execution of Algorithm MV_Creation

Step 1: Here the function RARM_Computation is executed to construct the table ranked attribute and relationship matrix (RARM) as shown in Table 6.

Step 2: View_construction is executed in this step to finally create the desirable materialized views. Say user specifies 2 materialized views V_1 and V_2 of size S_1 and S_2, respectively. V_1 would be created based on A_6 as it is the most important attribute of RARM. Now we need to check ARM to construct V_1. In the A_6 row of ARM it is found that A_1, A_3, A_4, A_7, A_9, and A_{10} all have the minimum deviation. Then we look at AQRM to check the Total_use. The value is found 6 for A_1, A_3, A_9, and A_{10}. In order to break the tie, we again check ARM to look at the Total_Deviation column. We find A_9 has the smallest value hence it is clubbed with A_6. If the size of V_1 is less than S_1, we proceed to include further attributes. Now according to the logic A_3 is the next best attribute and hence it is clubbed in V_1. If the size of V_1 is less then S_1, then we continue further and hence combine A_{10} with V_1 and again compare with S_1. Say at this point it is found the size of V_1 is more than S_1. Hence A_{10} is discarded and finally V_1 is constructed with A_6, A_9, and A_3.

Now the materialized view V_2 would be constructed. It would be based on A_9, as this one is the second most important attribute in RARM.

6   Performance

In this section a comparative study is presented between this methodology and the linear methodology of [17]. A software has been implemented using both the methodology. The implementation was carried out using .Net (visual Studio 2012) as front end and SQL Server 7.0 for the database and the operating system is Windows 7. For the experimental result we used an Intel core i3 3.1 GHz with the hard disk of 500 GB and RAM of 4 GB.

In Fig. 1, a comparative study is depicted based on the hit ratio of the proposed nonlinear methodology with [17]. The software was tested on a large student

Fig. 1 Comparative study of linear and nonlinear methodology

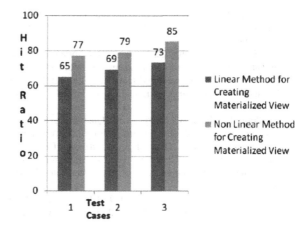

database. Three separate query sets were considered and each time it ran on the same database. Every test case showed better result in the proposed nonlinear methodology.

7 Conclusion

The materialized view creation process is based on attribute. Attribute is the most granular element in relational database. This methodology is designed such that it can be applied in any application domain irrespective of the size, as the number of views and their size are given by the users. The idea of using nonlinear methodology is a new concept in materialized view creation. Other nonlinear methodologies could be exercised further for better result.

The further extension of the work is exploiting materialized view maintenance schemes like full maintenance and incremental maintenance. When the performance of materialized view degrades, maintenance policy is required to avoid the construction of materialized view from scratch. This is an interesting area of research and can also be incorporated in this proposed work.

References

1. Yang, J., Karlapalem, K., Li, Q.: A framework for designing materialized views in data warehousing environment. In: Proceedings of 17th IEEE International Conference on Distributed Computing Systems, pp. 458–465 Maryland, USA, May 1997
2. Harinarayan, V., Rajaraman, A., Ullman, J.: Implementing data cubes efficiently. In: Proceedings of ACM SIGMOD International Conference on Management of Data, pp. 205–216. Montreal, Canada

3. Goldstein, J., Larson, P.: Optimizing queries using materialized views: a practical scalable solution. In: Proceedings of ACM-SIGMOD, International Confernece on Management of Data (2001)

4. Lin, Z., Yang, D., Song, G., Wang. T.: User-oriented Materialized View Selection 7th IEEE International Conference on Computer and Information Technology, pp. 133–138 (2007)

5. Zhou, J., Larson, P.-A., Goldstein, J., Ding, L.: Dynamic materialized views. In: 23rd International Conference on Data Engineering (ICDE), vol. 2 of 3, pp. 501–510, Istanbul, Turkey (2007)

6. Guodong, L., Shuai, W., Chang'an, L., Quanzhong, M.: A modifying strategy of group query based on materialized view. In: 3rd International Conference on Advanced Computer Theory and Engineering (ICACTE), vol. 5, pp. 381–384, China (2010)

7. Gong, A., Zhao, W.: Clustering-based dynamic materialized view selection algorithm. In: Fifth International Conference on Fuzzy Systems and Knowledge Discovery (FSKD), vol. 5, pp. 391–395 (2008)

8. Aouiche, K., Jouve, P.-E., Darmont, J.: Clustering-based materialized view selection in data warehouses. In: Proceedings ADBIS 2006, LNCS 4152, pp. 81–95 (2006)

9. Agarawal, S., Chaudhuri, S., Narasayya, V.: Automated selection of materialized views and indexes for SQL databases. In: Proceedings of 26th International Conference on Very Large Databases, Cairo, Egypt (2000)

10. Griffin, T., Libkin, L.: Incremental maintenance of views with duplicates. Materialized Views, pp. 191–207. MIT Press, Cambridge (1999)

11. Dhote, C.A., ALi, M.S.: Materialized view selection in data warehousing, In: Information Technology: New Generations, Third International Conference on, Information Technology: New Generations, Third International Conference on 2007, ITNG, pp. 843–847 (2007)

12. Talebian, S.H., Kareem, S.A.: Using genetic algorithm to select materialized views subject to dual constraints. In: International Conference on Signal Processing Systems, Singapore, pp. 633–38, May 2009

13. Horng et al., J.T.: Materialized view selection using genetic algorithms in a data warehouse system. In: Proceedings of the Congress on Evolutionary Computation, Washington, D.C., USA, July 1999

14. Sen, S., Chaki, N., Cortesi, A.: Optimal space and time complexity analysis on the lattice of cuboids using galois connections for data warehousing. In: Fourth International Conference on Computer Sciences and Convergence Information Technology (ICCIT 2009), pp. 1271–1275, Seoul, Korea (2009)

15. Larsen, P.A., Zhou, J.: Efficient maintenance of materialized outer-join views. In: 23rd International Conference on Data Engineering (ICDE 2007), Istanbul, Turkey, April 2007

16. Sen, S., Dutta, A., Cortesi, A., Chaki, N.: A new scale for attribute dependency in large database systems. In: 11th International Conference on Computer Information Systems and Industrial Management Applications (CISIM-2012), pp. 266–277, Venice, Italy (2012)

17. Ghosh, P., Sen, S., Chaki, N.: Materialized view construction using linear regression on attributes. In: 3rd International Conference on Emerging Applications of Information Technology (EAIT 2012), pp. 214–222, Kolkata, India (2012)

Author Index

© Springer India 2016
R. Chaki et al. (eds.), *Advanced Computing and Systems for Security*,
Advances in Intelligent Systems and Computing 395,
DOI 10.1007/978-81-322-2650-5

Printed in the United States
By Bookmasters